SYSTEMATICS OF MIDDLE AMERICAN MASTIFF BATS OF THE GENUS MOLOSSUS

Patricia G. Dolan

TEXAS TECH UNIVERSITY PRESS
1989

SPECIAL PUBLICATIONS, THE MUSEUM
TEXAS TECH UNIVERSITY
NUMBER 29

Series Editor
J. Knox Jones, Jr.

Published 22 December 1989

This book was set in 10 on 12 Baskerville and printed on acid-free paper that meets the guidelines for permanence and durability of the Committee on Production Guidelines for Book Longevity of the Council on Library Resources.

Special Publications of The Museum are numbered serially and published on an irregular basis. Institutions interested in exchanging publications should address the Exchange Librarian at Texas Tech University.

Library of Congress Cataloging-in-Publication Data

Dolan, Patricia G.
     Systematics of Middle American mastiff bats of the genus Molossus /
Patricia G. Dolan
     p.   cm.—(Special publications / The Museum, Texas Tech
University; no. 29)
     Bibliography: p.
     ISBN 0-89672-203-1 (alk. paper)
     1. Molossus—Central America—Classification.   I. Title.
II. Series: Special publications (Texas Tech University, Museum);
no. 29.
QL737.C54D65   1989
599.4—dc20   89-5096
                                                                              CIP

Texas Tech University Press
Lubbock, Texas 79409-1037, USA

CONTENTS

The family Molossidae occupies tropical and temperate zones in both the eastern and western hemispheres and has come to encompass a distinctive group of chiropterans that count among their most notable features a double articulation of the shoulder joint, a tail that projects conspicuously beyond the free edge of the uropatagium, and a narrow wing that results from the fifth digit being scarcely longer than the metacarpal of the first (Miller, 1907). Together with other refinements in skeletal structure, musculature, and general exophenotype, molossids are deemed highly adapted for rapid flight and maneuverability (Vaughan, 1978).

Although the family ranges worldwide, the first molossid to be described was a member of the endemic New World genus *Molossus* and was named *Vespertilio molossus* by Pallas in 1766. The specific epithet was inspired by the dog-faced appearance of the bat and its resemblance to a large black mastiff from the Greek Province of Molossis (Freeman, 1981). During the next 147 years, there was a veritable explosion in the number of named forms thought to occupy the Neotropics. Miller himself (1913) recognized no fewer than 19 specific taxa, many of these restricted to single islands of the Greater and Lesser Antilles. Multiple factors contributed to this proliferation of species names, not the least of which was the prevailing philosophical position of the early and middle 1800s that viewed species as immutable entities. Consequently, many authors of early descriptions of species felt little compulsion to refer to specific specimens, examine series, or provide more than a cursory allusion to place of origin. Thus, type specimens often were individuals of unknown sex with little label information and without special notation identifying them as types in the repositories in which they were housed. Even with the awakening concept of variation as a key element to be reckoned with in defining species, researchers still were hampered by improperly preserved, damaged, or lost type material, and were forced to make what sense they could of incomplete and occasionally inaccurate published records when attempting to relate new material to already described taxa.

Marked sexual dimorphism and unusually high degrees of local variation superimposed on a background of strong phenetic similarity continued to confound later attempts to identify species groupings and produced yet more taxa assigned to *Molossus* (J. A. Allen, 1916; Goodwin, 1956, 1959; Gardner, 1966). Genoways *et al.* (1981) recently demonstrated significant mensural differences between intraisland populations of *M. molossus* that substantiated claims by Jones *et al.* (1971) of extremely localized occurrences of this species, presumably resulting in genetically independent demes.

The last generic treatment was that of Miller (1913). With the addition of considerable material collected since that time and the application of sensitive statistical tests, it is now possible to examine patterns of geographic and non-geographic variation in morphology, to separate morphotypes, and to determine whether a model of pronounced mensural divergence among populations is typical of the genus as a whole and if so search for common causal relations. Because some geographic regions are poorly represented in collections, this study focused primarily on those portions of Central America north of the Canal Zone in Panamá. However, several extralimital samples were included in order to resolve as many nomenclatorial questions of synonymy and distribution as possible within the constraints imposed by limited geographic representation.

The ultimate delineation of a species depends on defining gene pools and establishing limits of genic exchange. In this study, electrophoretic data have been useful in determining species groupings by identifying and substantiating (by means of isoelectric focusing) fixed allelic differences in several of the morphotypes. Analyses of intraspecific variation in gene frequencies also have served to test independently the supposition that populations are isolates and subject to consequent levels of inbreeding sufficient to promote morphological and genetic divergence.

Advancements in chromosomal banding techniques during the past decade have provided a means whereby putative homologous segments can be identified and the number and types of rearrangements associated with the evolution of taxa traced. The reality of chromosomal variation, and especially the occurrence of karyotypic megaevolution, which is well documented by Baker and Bickham (1980) for the Chiroptera has spawned much debate and resulted in the formulation of two principal models. Wilson *et al.* (1975), Bush *et al.* (1977), and Lande (1979) maintain that small, inbred demes acting in concert with local bottlenecks and extinctions promote the fixation of new arrangements. The canalization model of Bickham and Baker (1979) attaches little weight to deme size and attributes chromosomal variation to the set of new selective pressures encountered by a lineage when it invades a novel niche that allows a greater number of rearrangements to exist at a selective advantage; geologically older groups exhibit fewer karyotypic differences as a result of selection for an optimum karyotype. Baker and Bickham (1980) have suggested the likelihood that karyotypic megaevolution also might be related to genetic and environmental factors that increase rates of chromosomal mutation or decrease crossing over. The importance of deme size and inbreeding in determining rates of chromosomal evaluation can be tested in part by studying mastiff bats. If morphological and genetic analyses describe populations of *Molossus* as isolates engaging in little genic exchange, then chromosomal variation is to be expected.

Unravelling the systematic relationships of *Molossus* has necessitated the use of a broad-based data set. Only after the patterns and limits of morphological, karyotypic, and genetic variation have been established over a wide geographic range has it been possible to gain sufficient insight to recognize interspecific norms and interpret departures from these standards associated with geographic and nongeographic variation. The systematic relationships described here for Central American *Molossus* appear reasonably sound. But the complete answer is still not in. South America is scantily represented in collections, and thus the number of species of mastiff bats and their distribution on that continent remain poorly understood. Conspecificity between the small *Molossus* inhabiting the Greater Antillean islands and those occupying the mainland and Lesser Antilles also has yet to be shown conclusively. Due to the complex patterns of intraspecific variation associated with localization and isolation, resolving these and other questions of molossid taxonomy will require future researchers to use multiple character states in the taxonomic descriptions of populations.

Materials and Methods

In 1977 and 1978, 618 *Molossus* were collected from México southward to Chepo in Panamá and used in a morphological analysis of species relationships. Tissue for electrophoresis was taken from 558 of these for a genic survey, and additional tissues from selected bats were stored for chromosomal analysis. Specimens were prepared either as standard museum study skins or were preserved in alcohol; in a few instances, only the skull was retained. All material is housed in The Museum, Texas Tech University (TTU).

A supplemental sample of 170 specimens from outside the primary study area was included in the multivariate analysis to help illuminate nomenclatorial questions and to clarify distributional limits, bringing the total to 788. The total multivariate sample represented 60 localities from seven Caribbean islands and 10 Central and South American countries. Also, 16 type specimens (measurements provided by D. C. Carter) were inserted into the data set as unique taxonomic units to determine their phenetic relationships to larger, geographically diverse samples of *Molossus*. Although more types than this exist, sufficient cranial and external measurements were unavailable for the remainder to permit their inclusion in a multivariate analysis. Finally, several localities sampled during the 1960s by D. C. Carter were revisited during the course of this work and yielded additional specimens that allowed me to investigate the interplay between time and morphological change. A key to the species recognized, sample size, population, and locality is presented in Table 1.

In the species accounts that follow, the total number of specimens examined for each taxon is given in parentheses. Populations representing collections made during this study are listed by number and indexed in Table 1. Other localities refer to specimens for which sufficient mensural data were available (courtesy of D. C. Carter) to allow confident assignment to a taxon. These are arranged alphabetically and are followed by the number examined from that place and the repository. Institutional abbreviations are: AMNH—American Museum of Natural History, New York; BMNH—British Museum (Natural History), London; CM—Carnegie Museum of Natural History, Pittsburgh; KU—Museum of Natural History, University of Kansas, Lawrence; LACM—Los Angeles County Museum of Natural History, Los Angeles; MNHN—Muséum National d'Histoire Naturelle, Paris; MSU—Michigan State University, The Museum, East Lansing; SMF—Natur-Museum und Forschungs-Institut Senckenberg, Frankfurt a.M.; TCWC—Texas Cooperative Wildlife Collection, Texas A&M University, College Station; USNM—National Museum of Natural History, Washington, D.C. Selected additional records also are included from the published literature.

Morphological Analysis

The following external and cranial measurements, 16 in number, were recorded for the 788 individuals included in the analysis: total length (TL), length of tail (LT), length of hind foot (HF), length of ear from notch (EAR), length of forearm (FA), length of metacarpal III (MET3), length of metacarpal IV (MET4), greatest length of skull (GLS), condylobasal length (CB), zygomatic breadth (ZB), breadth of braincase (BB), postorbital constriction (PC), depth of skull (DS), length of maxillary toothrow (MT), greatest breadth across molars (BM), greatest breadth across canines (BC). All cranial measurements as well as measurements of the forearm and metacarpals were made to the nearest 0.1 millimeter (mm.) with dial calipers. Other external dimensions were recorded directly from museum specimen labels. Specimens were considered to be adults only if the basioccipital-basisphenoid suture was visibly fused.

Individual, age, and secondary sexual variation were analyzed for specimens from México and Central America with the statistical analysis system (SAS) designed and implemented by Barr *et al.* (1976). Means were calculated for each character noted above and a one-way analysis of variance was used to test for differences between age classes and between sexes for each locality. Coefficients of variation (CV) were calculated to determine the extent of character variability.

Geographic variation was analyzed by means of univariate (mean, standard deviation, standard error) and multivariate statistics. To assess the degree of divergence among localities with all characters considered simultaneously, a multivariate analysis of variance (MANOVA) in SAS was used. This program provided weighted combinations of the measurements, which maximized the distinction among groups. Significant differences among groups were not assumed *a priori*. Four criteria (Hotelling-Lawley's Trace, Pillai's Trace, Wilke's Criterion, and Roy's Maximum Root Criterion) were used to test the hypothesis of no overall locality effect, that is, no significant morphological differences between or among samples. Characteristic roots and vectors were then extracted and mean canonical variates computed for each locality. New orthogonal axes, termed canonical variates, were constructed to extract the next best combination of characters to discriminate among samples. Characters with the least within-sample and greatest between-sample variation were emphasized. Each eigenvalue and its corresponding canonical variate (characteristic root) represent an identifiable fraction of the total variation. Sample means and individuals were plotted on those canonical variates that accounted for the greater fractions of total variation. The relative importance of each original variable to a particular canonical variate was computed by multiplying the vector variable coefficient (eigenvalue) by the mean value of the dependent variable, summing all variable values for a particular vector, and then calculating the percent relative influence (percent loading) of each variable per vector.

Chromosomal Analysis

Tissue samples from embryos or ears of adult *Molossus* were collected under sterile conditions in the field and stored in Ham's F-10 Nutrient Mixture supplemented with 20 percent fetal calf serum, 1.8 percent Penicillin-Streptomycin, and 0.9 percent Mycostatin suspension to combat fungal and

TABLE 1.—*Populations of* Molossus *examined. Sample size is given with males preceding females. Both genetic and morphological data were taken from populations 1 through 52; the two data sets for population 59 were collected on different specimens.*

Species	Population	N	Locality
M. rufus	1	7, 14	México. Nayarit: Río de Cañas
"	2	0, 4	México. Guerrero: 10 km. E Acapulco, Río de la Sabana
"	3	7, 1	México. Chiapas: Pijijiapan
			México. Chiapas: Huehuetán
"	4	8, 11	México. Chiapas: 6 km. E Cintalapa de Figueroa
"	5[a]	0, 0	México. Yucatán: Merida
"	6	3, 6	Guatemala. Santa Rose: 10 km. S, 14 km. E Chiquimulilla, Río Margarita
			El Salvador. Ahuachapan: Río San Francisco and Hwy. 2
"	7	15, 15	El Salvador. Cuscatlán: Suchitoto
"	8	13, 13	El Salvador. Sosonate: La Libertad
"	9	4, 11	El Salvador. San Miguel: Río San Antonio
"	10	7, 8	Honduras. Santa Barbara: Santa Barbara
"	11	13, 5	Honduras. Yoro: Santa Rita
"	12	3, 3	Honduras. Cortes: San Francisco de Yojoa
"	13	1, 1	Honduras. Yoro: Yoro
"	14[b]	3, 4	México. Oaxaca: Tehuantepec, San Blas
M. pretiosus	15	5, 18	Nicaragua. Boaco: 14 km. S Boaco, Los Cocos
"	16	2, 8	Costa Rica. Guanacaste: Liberia
M. bondae	17	4, 12	Nicaragua. Zelaya: Rama
"	18	14, 6	Costa Rica. Cartago: Turrialba
M. coibensis	19	17, 16	Panamá. Chiriquí: La Concepcíon
			Panamá. Chiriquí: Alanje
"	20	4, 16	Panamá. Veraguas: San Francisco
"	21	3, 8	Panamá. Los Santos: Los Santos
"	22	9, 15	Panamá. Panamá: Chepo
"	23	9, 7	Panamá. Panamá: Colón
M. sinaloae	24[a]	0, 0	México. Yucatán: Merida
"	25	0, 1	México. Chiapas: Pijijiapan
"	26	1, 2	México. Jalisco: El Grullo
"	27	6, 7	Honduras. Yoro: Santa Rita
			Honduras. Cortes: San Francisco de Yojoa
"	28	5, 15	Honduras. Yoro: Yoro
"	29	0, 2	Nicaragua. Zelaya: 4 km. W Rama
"	30	2, 0	Nicaragua. Matagalpa: 6 km. N El Tuma
"	31	11, 12	Costa Rica. Alajuela: Cariblanco
"	43	1, 5	Nicaragua. Zelaya: Rama
M. aztecus	32	0, 1	México. Jalisco: El Grullo
"	35	1, 1	Guatemala. Huehuetenango: Aguacatán
"	39	2, 2	Nicaragua. Matagalpa: 6 km. N El Tuma
M. molossus	33	3, 4	México. Oaxaca: Tehuantepec, San Blas 1977
"	34	1, 0	México. Chiapas: Huehuetán

TABLE 1.—*Continued.*

Species	Population	N	Locality
"	36a	3, 2	Guatemala. Santa Rosa: 10 km. S, 14 km. E Chiquimulilla, Río Margarita
"	36b	0, 1	El Salvador. Ahuachapán: Río San Francisco and Hwy. 2
"	37	8, 3	El Salvador. San Miguel: Río San Antonio
"	38	3, 1	El Salvador. Sosonate: La Libertad
"	40, 44	6, 16	Nicaragua. Rivas: 6 km. NE Rivas, San Jorge
"	41	4, 3	México. Oaxaca: Tehuantepec, San Blas 1978
"	42	0, 2	El Salvador. Cuscatlán: Suchitoto
"	46	0, 2	Panamá. Veraguas: San Francisco
M. coibensis	50	2, 0	Venezuela. Miranda: Guatopo National Park
M. molossus	51	1, 9	Venezuela. Guarico: 45 km. S Calabozo
"	52	0, 1	Venezuela. Guarico: 45 km. S Calabozo
"	53	21, 33	Perú. Loreto: 1 mi. SW Aguaytia
"	54	16, 12	Ecuador. Napo Pastaza: 4 mi. W Puyo, Shell-Mera
M. coibensis	56	2, 1	Perú. Huánuco: 19 mi. S Tingo María
M. molossus	57	0, 5	Perú. Huánuco: 2 mi. N Tingo María
M. aztecus lambi[d]	58	1, 1	México. Chiapas: 11 km. NW Escuintla
M. molossus	59	3, 0	Dominica. St. Paul: Antrim Valley
"	60	11, 11	Guadeloupe. Basse-Terre: 2 km. N Ballif
"	61	3, 8	Montserrat. St. Anthony: mouth of Belham River
"	62	2, 9	Trinidad. St. George: Maracas Valley
"	63	6, 0	Trinidad. Maracas Valley, San Rafael Estate
"	64	0, 1	Puerto Rico. El Verde Research Station
"	65	0, 2	Haiti. Dept. du Sud: 3 km. S Beaumont
"	66	3, 3	Jamaica
M. pretiosus	67	1, 0	Venezuela. Miranda: Guatopo National Park
M. pretiosus[c]	69	3, 3	Venezuela. Distrito Federal: La Guaira
*M. pretiosus**	68	1, 0	Venezuela. Distrito Federal: La Guaira
*M. barnesi**	71	0, 1	French Guiana. Cayenne
*M. longicaudatus**	73	0, 1	?Probably somewhere in Lesser Antilles
M. obscurus[d]	74	1, 0	Antilles. Martinique (by restriction—Husson, 1962)
*M. o. currentium**	75	1, 0	Argentina. Corrientes: Goya
M. rufus[d,e]	76	1, 0	French Guiana. Cayenne (by restriction—Miller, 1913)
*M. coibensis**	77	1, 0	Panamá. Coiba Island
*M. daulensis**	78	1, 0	Ecuador. Guyas: Daule
*M. pygmaeus**	79	0, 1	Curaçao
*M. fortis**	81	1, 0	Puerto Rico. Luguilla
*M. nigricans**	82	1, 0	México. Tepic: Acaponeta
*M. bondae**	83	0, 1	Colombia. Magdalena: 7 mi. E Santa Marta, Bonda
*M. debilis**	85	0, 1	West Indies. St. Kitts Island
*M. trinitatus**	86	1, 0	Trinidad. Belmont: Port-of-Spain

TABLE 1.—*Continued.*

Species	Population	N	Locality
		1960 Sampling	
M. rufus		10, 10	México. Sinaloa: Acaponeta
M. rufus		10, 8	El Salvador. Cuscatlán: Suchitoto
M. coibensis		10, 10	Panamá. Chiriquí: La Concepción
M. sinaloae		10, 10	Costa Rica. Alajuela: Cariblanco
M. aztecus		10, 10	Guatemala. Huehuetenango: Aguacatán

[a] Only forearm measurements available, therefore excluded from MANOVA.

[b] topotype.

[c] paratype.

[d] syntype.

[e] lectotype.

[*] holotype.

bacterial contamination. In the laboratory, fibroblast cultures begun from these tissues were maintained in the above media, without Mycostatin, at 35°C. Dividing cells were arrested at metaphase by applying 0.1 to 0.5 milliliters of 0.0005 percent Velban in 15 milliliters of media for 20 minutes and then were harvested with 0.25 percent trypsin. Karyotypes were prepared and G-bands produced as outlined by Greenbaum *et al.* (1978). Procedures for C-banding follow those of Baker and Bass (1979).

The most frequently encountered chromosome number in counting 10 or more spreads on a single slide was taken as the diploid number (2N) for that species. Terminology describing centromeric placement is according to Patton (1967).

Electrophoretic Analysis

Heart and kidney extracts collected in the field were stored together and frozen in liquid nitrogen; liver biopsies were frozen separately. Tissue was prepared for electrophoresis by masceration in approximately five milliliters of a stock grinding solution (0.01M Tris-0.001M EDTA, pH 6.8; four milliliters 0.01M NADP stock solution) and spun in a refrigerated centrifuge for 20 minutes. The supernatant collected was frozen for later use. Instructions for the preparation of buffers and biochemical stains were taken from Selander *et al.* (1971). A summary of the systems examined in this study is given in Table 2 together with experimental conditions.

Allelic designation follows Greenbaum (1978) with the most common allele at a locus being 100, if migration is anodal, or −100, if movement is cathodal. Other alleles are described in terms of percentage migration relative to the 100 (−100) allele. When more than one locus was found within a system, the locus with the greatest mobility was designated "1" and progressively slower loci received increasing numerical values. Allelic differences were confirmed by serial side-by-side comparisons.

TABLE 2.—*Summary of test conditions and materials used in the electrophoretic analysis of* Molossus.

Gel type	Tissue	Milli-amperage (ma.)	Time (hrs.)	Stains
Continuous Tris-Citrate I pH 8.0	Liver	75	6	Esterase (EST) αGlycerophosphate dehydrogenase (α-GPD) Alcohol dehydrogenase (ADH) Glutamic oxalacetic transaminase (GOT) Albumin (ALB) General Protein (GP) Sorbitol dehydrogenase (SDH)
Continuous Tris-Citrate II pH 6.7 (gel) pH 6.3 (tray)	Kidney	75	4	Malate dehydrogenase (MDH) Lactate dehydrogenase (LDH) Isocitrate dehydrogenase (IDH) Hemoglobin (Hb)
	Liver	50	5–6	Phosphoglucomutase (PGM) Glutamate dehydrogenase (GDH)
Tris EDTA Borate pH 8.8	Kidney		7–8	Indophenol oxidase (IPO) Glucose-6-phosphate dehydrogenase (G6P)

Estimates of genetic variability within and among populations were obtained by means of Wright's (1965) *F*-statistics as modified by Nei (1977) using a computer program developed by R. K. Chesser. A chi-square test was used to identify significant heterogeneity in allele frequencies among molossid populations.

Coefficients of genetic similarity (Rogers' *S*) and distance (Nei's *D*) were calculated for all pair-wise comparisons of populations. *D* is a measure of the accumulated number of recognizable codon differences per locus (Nei, 1972), and *S* is based on the sum of geometric differences between allelic frequencies for each locus of every pair of populations being compared (Rogers, 1972). A cluster analysis was performed on the genetic similarity matrix of Rogers' *S* values using the unweighted pair-group method with arithmetic averages (UPGMA) provided in the NTSYS computer package (Rohlf and Kispaugh, 1972) to obtain a phenogram of population relationships.

Isoelectric Focusing

Vertical slab equilibrium polyacrylamide gels were run on critical loci (α-GPD, EST-2) to confirm similarities and differences. Gels, with total volume of 36 milliliters, consisted of: 1.75 milliliters pH 3-10, 40 percent ampholyte; 9.0 milliliters 30 percent acrylamide, 0.8 percent Bis stock solution; 2.1 milliliters Riboflavin-TEMED stock; 23.15 milliliters water. Weak UV exposure for 1.5–2.0 hours was used for polymerization. Samples for testing were mixed in a one-to-one ratio of 50 percent sucrose, eight percent carrier ampholyte mixture. Aliquots of approximately 20 microliters were overlayed with a 10 percent sucrose, eight percent carrier ampholyte solution. Gels were run 24 hours at 300V with a 0.02M NaOH cathode buffer and a 0.01M PO_4 anode buffer. Biochemical stains were the same as those used for the electrophoretic analysis. Stained gels were fixed in a methanol, acetic acid, and water mixture overnight, photographed, and dried. Isoelectric points were determined by measuring migration distance of bands from the wells and plotting those values against pH gradients taken from the gel. Quantitative counts of band number and intensity were made with a densitometer.

RESULTS

Nongeographic Variation

Sexual dimorphism.—The number of adult males and females, respectively, included in a one-way ANOVA to test for sexual variation within each species are given in parentheses followed by those characters out of the 16 examined not showing variation due to sex at a probability level of *P* less than or equal to 0.05: *M. rufus* (43, 66) HF; *M. pretiosus* (7, 27) DS, HF, LT; *M. bondae* (18, 18) GLS, ZB, DS, LT; *M. sinaloae* (26, 44) BB; *M. coibensis* (43, 66) all characters significantly different; *M. molossus* (30, 37) PC, DS, HF, EAR, FA, MET3, MET4; *M. aztecus* (3, 4) PC, DS, HF, EAR, LT, FA, MET3, MET4. A key to abbreviations appears in the section on materials and methods. Data reported here are a composite of all populations within a species factored only by sex, but identical trends were evinced when populations were analyzed separately. Males averaged larger than females in almost every respect in all species, but sexual differences were especially evident when comparing cranial features. Among the smaller taxa, *M. aztecus* and *M. molossus,* there was a marked convergence in size that tended to obscure the sexual differences of external characters. Representative cranial and external measurements are provided in Tables 3 and 4 for both sexes in geographically selected populations of each Central American species recognized in this work. As a consequence of discernible size variation between sexes, males and females were not pooled in subsequent MANOVAs.

Age variation.—Incomplete fusion of the basioccipital-basisphenoid suture and little-to-no toothwear were characters used to identify juveniles. Of the 618 Central American specimens collected, only 40 were classified as such, and most of these were members of the species *rufus.* For this reason, only *M. rufus* was used to test for morphological differences between age classes. Populations were pooled but partitioned by sex and age class within sex.

A one-way classification ANOVA revealed an interesting difference within sexes. The number of characters significantly different at the *P* less than or equal to 0.05 level together with the minimum number of adults and juveniles sampled, respectively, were found to be: males (85, 8) GLS, ZB, PC, DS, CB, BC, TL, HF, EAR; females (108, 7) GLS, ZB, TL. A greater disparity in size was noted between adult and nonadult males than among females, suggesting, perhaps, a more rapid maturation rate among the latter. Because there was detectable interclass size variation, juveniles were excluded from additional morphological analyses. However, their tissue samples were included in the genetic portion of this study.

Data gathered here extend the information on the reproductive strategy of polyestry described by Carter (1970) for two species of *Molossus* to five, and in all likelihood it is the inherent pattern ascribed to by all members of the genus. Females of all taxa, with the exception of *M. sinaloae* and *M. aztecus,* were taken that were simultaneously lactating and carrying embryos. Given the polyestrous habit of mastiff bats, the dearth of young (40) encountered

TABLE 3.—*Selected cranial measurements for seven species of Molossus in Central America and México. Means are followed by range, in parentheses, and one standard deviation about the mean. Where sample size varies, it is noted in italics. See text for key to populations.*

Population no., sex, and sample size	Greatest length of skull	Condylobasal length	Breadth of braincase	Maxillary toothrow length	Breadth across M3-M3	Breadth across canines
Molossus rufus						
1. males, 7	23.9(22.8–24.5)0.58	21.2(19.7–21.8)0.69	11.3(10.8–11.7)0.26	8.6(8.2–8.8)0.20	10.4(9.7–10.6)0.33	6.4(6.1–6.5)0.14
females, 18	22.9(22.6–23.6)0.26	20.5(20.0–21.1)0.26	11.0(10.6–11.3)0.21	8.3(8.2–8.6)0.14	10.2(9.9–10.4)0.15	6.0(5.8–6.2)0.09
2. females, 4	21.6(21.4–21.7)0.14	19.2(19.0–19.4)0.16	10.4(10.3–10.5)0.10	7.8(7.7–7.8)0.06	9.3(9.3–9.4)0.05	5.5(5.4–5.6)0.08
3. males, 8	23.0(22.3–24.2)0.60	20.31(19.6–21.1)0.49	11.1(10.8–11.6)0.29	8.1(7.9–8.5)0.21	9.8(9.3–10.4)0.38	6.0(5.8–6.1)0.10,7
females, 3	21.6(21.1–21.8)0.40	20.1(18.8–19.6)0.40	11.0(10.9–11.1)0.12	7.7(7.5–7.9)0.21	9.2(9.1–9.3)0.10	5.3(0)0.1
4. males, 9	24.0(23.2–25.0)0.51	21.1(20.6–21.6)0.31	11.2(10.8–11.9)0.36	8.3(8.1–8.5)0.13	10.3(10.0–10.5)0.17	6.2(5.9–6.5)0.22
females, 19	22.5(21.5–23.1)0.36	20.1(19.4–20.8)0.35	10.9(10.5–11.3)0.22	8.0(7.5–8.4)0.19	9.9(9.2–10.2)0.27	5.8(5.6–6.1)0.16
7. males, 15	23.3(22.9–23.7)0.24	20.6(20.1–20.9)0.21	11.0(10.6–11.4)0.27	8.1(7.8–8.3)0.15	9.9(9.5–10.2)0.23	6.0(5.6–6.3)0.16
females, 15	21.8(21.2–22.6)0.47	19.6(19.1–20.1)0.28	10.8(10.3–11.3)0.25	7.8(7.2–8.1)0.22	9.5(9.2–9.8)0.15	5.6(5.4–5.8)0.15
11. males, 13	23.8(23.3–24.3)0.31	21.2(20.5–21.6)0.28	11.5(11.1–11.7)0.19	8.4(8.1–8.6)0.15	10.2(10.0–10.6)0.19	6.2(5.9–6.5)0.17
females, 6	22.9(22.2–23.3)0.44	20.3(19.6–20.8)0.43	11.1(10.8–11.3)0.18	8.1(7.8–8.2)0.17,5	9.9(9.4–10.2)0.30	5.8(5.5–6.2)0.24
14. males, 3	23.2(22.7–23.8)0.57	20.3(19.7–21.0)0.65	11.0(10.8–11.2)0.20	8.2(8.2–8.3)0.06	10.2(10.0–10.4)0.20	6.0(5.9–6.2)0.17
females, 4	22.0(21.4–22.4)0.53	19.7(19.4–20.0)0.28	10.7(10.6–10.8)0.10	8.1(8.0–8.2)0.10	9.9(9.6–10.1)0.24	5.8(5.5–6.0)0.21
Molossus pretiosus						
15. males, 5	21.7(21.5–22.0)0.19	19.2(19.0–19.4)0.16	10.3(10.1–10.6)0.20	7.3(7.1–7.5)0.18	9.3(9.1–9.6)0.22	5.6(5.5–5.8)0.13
females, 18	20.2(18.8–20.8)0.47	18.0(16.4–18.6)0.49	10.1(9.7–10.6)0.26	7.0(6.3–7.4)0.24	9.0(8.5–9.3)0.24	5.2(4.8–5.5)0.16
16. males, 3	21.7(21.4–21.9)0.25	18.6(18.2–18.9)0.38	10.5(10.3–10.7)0.21	7.2(7.2–7.2)0.06	9.2(8.9–9.4)0.26	5.7(0)0.2
females, 8	20.2(19.7–20.9)0.35	17.8(17.5–18.1)0.18	10.1(9.6–10.6)0.28	7.0(6.8–7.1)0.11	8.7(8.3–9.0)0.24	5.2(5.0–5.3)0.10
Molossus bondae						
17. males, 4	20.6(20.3–20.9)0.28	17.9(17.5–18.1)0.26	10.1(9.7–10.3)0.27	6.9(6.6–7.1)0.21	8.8(8.5–9.0)0.22	5.3(5.1–5.4)0.15
females, 12	18.8(18.3–19.9)0.49	16.6(16.0–17.8)0.49	9.7(9.3–10.1)0.24	6.6(6.3–7.0)0.21	8.4(7.9–9.0)0.28	4.7(4.4–4.9)0.14
18. males, 14	20.1(19.8–20.5)0.20	17.6(17.3–18.0)0.20	10.0(9.7–10.4)0.19	6.8(6.4–7.0)0.15	8.7(8.5–9.0)0.16	5.1(5.0–5.3)0.09
females, 6	18.8(18.1–19.4)0.44	16.6(16.0–17.2)0.42	9.5(9.2–10.0)0.31	6.4(6.4–6.5)0.05	8.5(8.2–8.6)0.15	4.7(4.6–4.8)0.10
Molossus coibensis						
19. males, 19	17.7(17.2–18.0)0.27	15.5(15.1–15.8)0.17	9.1(8.8–9.5)0.15	6.2(5.9–6.4)0.13	8.0(7.7–8.2)0.16,18	4.7(4.4–4.8)0.12,*18*
females, 16	16.7(16.4–17.1)0.19	14.9(14.7–15.2)0.15	8.9(8.7–9.1)0.10	5.9(5.7–6.1)0.12	7.7(7.3–7.9)0.19	4.2(4.1–4.4)0.08
22. males, 10	17.0(15.9–18.4)0.62	15.1(14.2–16.0)0.45	8.8(8.6–9.1)0.19	5.9(5.6–6.2)0.17	7.6(7.3–7.9)0.15,*9*	4.4(4.1–5.8)0.20
females, 15	16.2(15.6–16.8)0.29	14.4(13.8–14.8)0.26	8.7(8.2–8.9)0.18	5.6(5.4–5.8)0.11	7.4(7.1–7.8)0.21	4.1(3.9–4.4)0.14

TABLE 3.—*Continued.*

Population no., sex, and sample size	Greatest length of skull	Condylobasal length	Breadth of braincase	Maxillary toothrow length	Breadth across M3-M3	Breadth across canines
Molossus sinaloae						
26. male, 1	22.3	20.2	10.1	8.1	9.7	5.7
females, 2	21.6(0)0	19.3(0)0	10.0(9.9–10.1)0.14	7.9(0)0	9.3(9.1–9.4)0.21	5.4(5.3–5.4)0.07
28. males, 5	20.9(20.5–21.8)0.54	18.6(18.4–18.8)0.16	9.9(9.7–10.5)0.35	7.2(7.2–7.3)0.05	8.9(8.6–9.1)0.21	5.3(5.1–5.4)0.11
females, 16	19.8(19.2–20.8)0.42	17.7(17.3–18.4)0.29	9.6(9.3–10.2)0.23	7.0(6.8–7.3)0.13	8.5(8.2–9.0)0.17	5.0(4.8–5.2)0.12
31. males, 11	21.8(21.0–22.6)0.40	19.5(18.8–20.2)0.38	10.0(9.8–10.3)0.15	7.6(7.4–7.9)0.16	9.1(8.9–9.4)0.17	5.5(5.3–5.7)0.12
females, 12	20.3(19.9–20.8)0.31	18.3(17.8–18.7)0.27	9.7(9.5–10.1)0.19	7.3(7.1–7.4)0.10	8.7(8.4–8.9)0.13	5.1(5.0–5.2)0.09
Molossus aztecus						
32. female, 1	17.2	15.4	8.9	5.9	7.9	4.4
35. male, 1	18.5	16.6	9.4	6.5	8.4	4.7
female, 1	17.2	15.4	8.9	6.1	7.8	4.3
39. males, 2	18.7(18.4–19.0)0.42	16.5(16.4–16.6)0.14	9.5(9.4–9.5)0.07	6.6(6.5–6.6)0.07	8.4(8.3–8.5)0.14	5.0(4.9–5.0)0.07
females, 2	17.5(0)0	15.6(0)0	9.3(9.2–9.3)0.07	6.2(6.1–6.2)0.07	8.1(8.0–8.1)0.07	4.5(4.4–4.5)0.07
Molossus molossus						
33. males, 3	17.5(17.3–17.8)0.26	16.0(15.9–16.1)0.12	9.1(8.9–9.2)0.15	6.1(6.0–6.3)0.15	7.7(7.4–7.9)0.25	4.5(4.3–4.8)0.25
females, 4	16.7(16.4–17.0)0.28	15.1(15.0–15.2)0.12	8.8(8.6–9.1)0.24	5.9(5.8–6.0)0.10	7.5(7.2–7.8)0.28	4.2(4.0–4.3)0.13
36. males, 3	17.1(16.8–17.4)0.31	15.2(14.6–15.6)0.55	8.8(8.7–8.9)0.12	6.0(5.9–6.2)0.15	7.6(7.2–7.8)0.35	4.4(4.3–4.5)0.10
females, 4	16.4(15.8–16.8)0.43	14.9(14.3–15.2)0.39	8.5(8.3–8.8)0.22	5.9(5.7–6.1)0.17	7.2(7.1–7.4)0.14	4.1(3.9–4.2)0.14
37. males 8	17.5(17.1–17.7)0.19	15.7(15.4–15.9)0.20	8.7(8.5–8.8)0.11	6.2(6.0–6.3)0.10	7.6(7.5–7.8)0.10	4.5(4.3–4.7)0.14
females, 3	16.6(16.2–17.2)0.51	15.0(14.4–15.5)0.55	8.4(8.2–8.8)0.32	6.0(5.9–6.1)0.12	7.4(7.2–7.5)0.15	4.1(4.1–4.2)0.06
42. males, 6	18.1(17.7–18.4)0.25	16.2(15.9–16.5)0.25	9.2(9.0–9.4)0.18	6.3(6.2–6.4)0.08	7.9(7.7–8.0)0.14	4.7(4.6–4.8)0.08
females, 16	17.3(16.3–17.7)0.29	15.5(14.5–15.8)0.33	8.9(8.4–9.1)0.18	6.1(5.8–6.3)0.12	7.5(7.2–7.9)0.17	4.3(4.0–4.5)0.13

TABLE 4.—*Selected external measurements for seven species of* Molossus *in Central America and México. Means are followed by range, in parentheses, and one standard deviation about the mean. Where sample size varies, it is noted in italics. See text for key to populations.*

Population no., sex, and sample size	Total length	Length of tail	Length of ear	Length of forearm	Length of metacarpal III	Length of metacarpal IV
Molossus rufus						
1. males, 7	134.0(118–140)7.3	45.9(37–50)4.2	18.7(17–19)0.8	52.8(47.8–54.4)2.3	54.3(48.5–56.5)2.8	52.4(47.2–54.6)2.6
females, 18	131.9(129–136)2.1	45.2(42–47)1.4	18.1(18–19)0.3	51.9(50.5–53.7)1.0	53.3(51.8–55.4)1.0	51.8(50.1–53.6)1.0
2. males, 4	121.5(118–124)3.0	41.5(40–44)1.9	16.3(16–17)0.5	48.2(47.7–49.1)0.6	48.9(48.7–49.1)0.2	47.7(47.2–48.1)0.4
3. males, 8	128.8(124–135)4.3	43.9(42–46)1.4	17.8(17–19)0.7	48.6(47.4–50.4)0.9	49.4(47.8–50.7)0.8	47.9(46.3–49.1)1.0
females, 3	124.3(122–127)2.5	42.3(42–43)0.6	17.0(0)0	48.6(48.5–48.7)0.1	49.5(48.3–50.2)1.0	48.4(47.4–49.6)1.1
4. males, 9	137.2(135–141)2.7	46.3(43–49)2.4	17.7(17–19)0.7	51.5(49.1–53.4)1.5	52.8(50.9–54.7)1.2	51.4(49.8–53.6)1.2
females, 19	130.3(124–136)3.5	44.8(41–49)2.4	17.3(17–18)0.5	50.3(49.0–52.4)0.8	51.5(50.9–52.7)1.0,*11*	50.3(48.6–52.4)1.2,*11*
5. males, 3				53.0(49.0–52.4)0.8	54.3(53.5–55.2)0.9	52.6(52.0–53.5)0.8
females, 6				50.6(47.0–51.6)1.8	51.6(46.8–53.4)2.4	49.7(43.9–52.1)3.0
7. males, 15	129.3(123–136)3.7	46.2(41–50)2.1	17.1(16–18)0.5	51.3(50.1–52.1)0.6	52.3(50.7–53.7)0.9	50.9(49.5–52.5)0.9
females, 15	124.9(119–131)4.2	44.2(40–47)1.6	17.3(16–18)0.6	50.0(48.3–51.6)1.1	50.9(48.7–52.9)1.2	49.6(47.1–51.5)1.2
11. males, 13	133.0(128–140)3.2	47.3(44–51)2.5	18.3(18–19)0.5	52.8(52.2–53.3)0.3	54.0(53.4–55.7)0.6	52.2(51.8–53.2)0.4
females, 6	126.8(122–134)4.8	44.3(41–48)2.5	17.5(17–18)0.5	51.2(48.9–52.4)1.3	52.1(50.0–53.6)1.3	50.6(49.0–51.8)1.0
14. males, 3	129.7(125–133)4.2	45.3(43–47)2.1	18.3(18–19)0.6	50.0(48.0–51.4)1.8	51.0(49.4–52.9)1.8	49.7(48.0–51.4)1.7
females, 4	125.5(121–128)3.1	44.0(42–46)1.6	17.0(16–18)0.8	49.2(48.6–50.0)0.6	49.7(48.8–50.2)0.6	48.4(47.6–49.1)0.6
Molossus pretiosus						
15. males, 4	119.0(115–122)2.9	43.5(43–44)0.6	16.6(16–17)0.5	45.8(44.0–47.8)1.3	47.2(45.8–49.9)1.0	45.9(43.9–47.9)1.3
females, 23	113.1(111–117)1.8,*17*	42.2(38–46)2.1,*17*	16.1(15–17)0.7,*18*	44.5(41.6–45.9)0.9	46.0(43.1–47.9)1.1	44.7(42.1–46.7)1.1
16. males, 3	115.7(115–116)0.6	40.7(39–42)1.5	17.0(0)0	44.7(43.3–46.5)1.6	47.1(46.7–47.8)0.6	45.6(45.2–46.2)0.5
females, 10	111.2(108–115)2.6	39.2(36–43)1.9	16.1(15–17)0.6	44.4(43.4–46.0)0.8	45.7(44.6–48.5)1.2	44.3(43.3–46.6)1.0
Molossus bondae						
17. males, 4	111.0(109–112)1.4	40.5(40–41)0.6	15.5(15–16)0.6	41.6(41.3–42.0)0.3	43.1(42.2–43.8)0.7	41.9(41.0–42.4)0.7
females, 12	103.3(100–109)3.2	37.8(35–39)1.1	14.7(14–15)0.5	40.1(38.6–41.2)0.8	41.6(39.7–42.5)1.0	40.6(38.9–41.4)0.8
18. males, 14	110.6(106–114)2.6	40.0(36–42)1.8	15.9(15–16)0.3	41.4(40.3–42.8)0.8	43.3(42.1–44.7)0.8	42.0(40.9–43.7)0.9
females, 6	106.0(102–110)3.2	38.8(36–40)1.6	13.9(13–16)0.5	40.5(38.4–41.5)1.1	42.1(39.5–43.9)1.4	41.1(38.7–43.0)1.4

TABLE 4.—*Continued.*

Population no., sex, and sample size	Total length	Length of tail	Length of ear	Length of forearm	Length of metacarpal III	Length of metacarpal IV
			Molossus coibensis			
19. males, 20	96.4(93–101)2.6,19	34.6(31–37)1.7,18	13.9(13–15)0.7,19	36.0(34.8–36.8)0.6	37.9(36.9–39.1)0.6	36.7(35.7–38.0)0.6
females, 23	90.7(86–97)2.6,17	31.8(28–34)1.8,16	13.5(13–14)0.5,17	34.7(33.6–35.6)0.5	36.5(35.2–37.8)0.6	35.5(34.2–36.5)0.6
22. males, 11	89.9(82–95)3.6,10	32.1(30–35)1.7,9	13.8(13–14)0.4,10	35.0(34.1–35.7)0.5	36.5(35.6–37.4)0.6	35.1(34.4–36.1)0.6
females, 20	88.4(85–92)2.2,15	31.1(29–33)1.0,15	13.5(13–14)0.5,15	33.9(32.6–34.9)0.7	35.6(33.6–36.7)0.8	34.4(32.4–35.4)0.8
			Molossus sinaloae			
24. males, 11				46.3(44.6–47.0)0.9	48.1(46.3–49.2)0.9	46.3(44.7–47.1)0.8
females, 2				46.0(45.2–40.8)1.1	48.0(47.5–48.5)0.7	46.6(46.1–47.0)0.6
26. male, 1	141.0	53.0	17.0	51.5	54.3	52.8
females, 2	129.5(128–131)2.1	47.5(46–49)2.1	16.0(0)0	49.5(49.4–49.6)0.1	51.3(50.8–51.8)0.7	50.1(49.9–50.3)0.3
28. males, 7	120.0(115–124)3.4	45.6(44–46)0.8	15.0(0)0	47.1(46.0–47.9)0.7	48.6(46.7–50.6)1.4	47.1(45.4–49.0)1.3
females, 16	115.6(109–120)3.2	43.4(41–45)1.2	15.0(0)0	47.1(45.4–48.4)0.9	48.7(46.7–50.6)1.2	47.0(44.3–49.0)1.3
31. males, 11	127.9(121–135)4.5	46.5(42–51)2.5	15.8(15–16)0.4	49.3(47.6–51.1)1.0	51.1(49.6–52.8)0.9	49.0(47.3–50.8)0.9
females, 13	122.6(117–128)3.1	43.4(40–46)1.9	14.9(14–15)0.3	47.8(46.7–49.0)0.6	50.1(48.8–51.0)0.7	48.2(47.0–49.3)0.7
			Molossus aztecus			
32. female, 1	98.0	31.0	14.0	36.6	39.0	37.6
35. male, 1	103.0	39.0	15.0	37.8	40.2	38.0
female, 1	98.0	34.0	15.0	37.7	40.0	38.2
			Molossus molossus			
39. males, 2	99.5(99–100)0.7	36.0(35–37)1.4	14.5(14–15)0.7	38.1(37.7–38.4)0.5	40.2(39.6–40.7)0.8	38.7(38.0–39.4)1.0
females, 2	96.0(95–97)1.4	34.5(34–35)0.7	14.0(0)0	37.4(36.9–37.8)0.6	39.2(0)0	37.9(37.8–37.9)0.1
33. males, 3	100.3(99–103)2.3	35.7(35–36)0.6	14.0(0)0	38.9(38.5–39.5)0.5	41.3(40.6–41.7)0.6	40.6(39.9–41.1)0.6
females, 4	96.5(94–98)1.9	34.0(33–36)1.4	13.3(13–14)0.5	38.1(37.3–38.7)0.7	39.9(38.2–40.9)1.2	39.1(34.5–39.7)1.1
36. males, 3	93.0(92–94)1.0	33.0(32–35)2.0	13.0(0)0	37.8(37.5–38.2)0.4	39.5(39.3–39.9)0.3	38.0(37.4–38.3)0.5
females, 4	93.0(91–96)2.2	33.8(32–35)1.3	13.3(13–14)0.5	37.1(35.8–38.1)1.2	38.2(36.3–40.6)2.2	37.3(35.8–39.3)1.8
37. males, 8	100.6(96–104)2.6	35.5(34–38)1.3	13.3(13–14)0.5	38.3(37.1–39.8)0.9	40.2(39.0–42.2)1.0	38.9(37.6–40.5)0.9
females, 3	95.3(91–98)3.8	32.3(30–34)2.1	13.0(12–14)1.0	37.0(35.7–37.8)1.1	39.3(38.2–40.2)1.0	38.2(37.0–39.0)1.1
42. males, 6	101.0(98–105)2.6	37.2(35–39)1.6	15.0(0)0	39.5(38.8–40.1)0.4	40.7(39.0–41.7)0.9	39.6(38.9–40.4)0.6
females, 17	97.2(90–100)2.5	35.7(33–39)1.6,16	14.3(14–15)0.5	39.1(37.9–40.1)0.6	41.0(39.4–42.2)0.8	39.6(38.3–41.0)0.8

during my midsummer field season (late June to middle August) is perplexing. Data in hand, assuming a 90-day gestation period is typical of Neotropical representatives of the family (Carter, 1970), point to two peaks in parturition (April–May and July–August). The apparent absence of spring-born young at midsummer could signify high infant mortality or such rapid physical development that juveniles quickly blend into the population—at birth, newborns are already 25 percent of adult weight. Field observations reveal that juveniles and subadults cohabit parental roost sites.

Individual variation.—Coefficients of variation (CV) were averaged for populations within species after cranial and external characters were separated. CVs, by taxon, are given with cranial averages preceding those of external variates, and values for males preceding those for females. The number of populations in each set appears in parentheses: *M. rufus* 2.3, 2.8 (12); 2.2, 2.6 (13); *M. pretiosus* 2.0, 2.5 (2); 2.8, 3.0 (2); *M. bondae* 2.1, 2.6 (2); 2.7, 3.1 (2); *M. coibensis* 2.3, 2.9 (5); 2.3, 3.1 (5); *M. sinaloae* 2.3, 2.6 (5); 1.9, 2.4 (7); *M. molossus* 2.2, 3.0 (6); 2.3, 3.3 (6); *M. aztecus* 2.1, 2.6 (1); 1.6, 1.4 (1). The one truism evident from the foregoing is the inherently higher degree of variability associated with external characters (see Long, 1968), a factor I attribute to difficulty in obtaining precise measurements. With this in mind, external characters were deleted from ensuing multivariate analyses. The most variable of the 16 characters treated were length of hind foot (CV 0 to 18.3) and tail (CV 0 to 13.6), but CVs for ear length were also broadly distributed.

No obvious geographic trends in variability existed among populations, nor were species-related differences in degree of morphological variation evident. Although some authors (for example, Yates and Schmidly, 1977) have observed higher average CVs for males than for females, members of the genus *Molossus* reaffirm Long's (1969:298) conviction that in general "there is as yet no significant basis for attributing greater variability to one sex in mammals as a group."

Relative to other members of the class Mammalia, Long (1968) observed that the Chiroptera typically exhibited low variation. Using his CV values for cranial length (1.5–2.6) as a baseline, it is apparent that the family Molossidae conforms to this basic chiropteran theme: *Eumops* (Eger, 1977) CV 1.5–2.8; New World *Tadarida* (Carter, 1962) CV 1.0–2.6, (Long and Jones, 1966) CV 2.1; African *Tadarida* (Peterson, 1971, 1974) CV 0.9–2.1; *Molossus* data presented herein.

Coloration.—Unless stated otherwise, the following descriptions pertain exclusively to Central American *Molossus*.

In terms of pelage characteristics, *M. rufus*, *M. pretiosus*, and *M. bondae* proved to be similar. Dorsal hairs were black to blackish in color, albeit somewhat paler at the base, and between 2.0 and 2.5 millimeters in length. The venter was always slightly paler than the dorsum. Additionally, the membranes, muzzle, and ears were of the same color as the fur.

Although Miller (1913) referred to two color phases (reddish and blackish) in *M. rufus* and *M. pretiosus*, neither of these species is truly dichromatic for there is a distinct progression from black, through deep russet, to ocher red in series of specimens examined. The transition from black to orangish red is presumably related to degradation of the melanistic medullary pigment granules as the hair becomes worn, which permits exposure of the underlying xanthophylls in the cortex. Geographic variation in color was not detected.

M. coibensis resembled the foregoing *rufus*-complex in most features of the pelage including color, which was described by Miller (1913:92) as "between the burnt-umber and seal-brown of Ridgway." However, it differed in showing a pale band of cream or white at the base of the dorsal hairs and in lacking the extreme ocher red color phase. As the pelage of this species became worn, it acquired a more brownish hue.

J. A. Allen (1906) referred to the type of *M. sinaloae* as dull dark brown above and much paler beneath. It, like *aztecus* and *molossus*, also has a well-defined basal white band on the dorsal hairs. Material examined from México and Middle America conformed to Allen's description when the pelage was fresh; worn hairs imparted a slight, reddish tinge to the fur, but development of the orangish red color never was as pronounced as in members of the *rufus* complex.

Few specimens of *M. aztecus* have been identified so that generalizations regarding pelage difference between it and *M. molossus* must necessarily be regarded as tenuous. The point to emphasize is that the two taxa are extremely similar externally. Nonetheless, the membranes, muzzle, and ears of *aztecus* appear blacker, the pelage deeper brown, and the white band along the base of dorsal hairs less conspicuous.

Central American *M. molossus* are a toffee brown, noticeably paler and duller than conspecifics from the Lesser Antilles, which are almost ebony. Material from South America was variable: specimens from Perú and Ecuador approached insular populations in their blackish color; those from Suriname approximated Middle American *M. molossus*, and specimens from Venezuela and Argentina were so pale as to be termed fawn-colored. With only limited comparative material at hand, it is likely that this perceptible variation merely reflects poorly understood patterns of individual or seasonal influences. On the other hand, it could be linked to distinct geographic areas and hence serve as a useful subspecific trait.

Geographic Variation

Univariate analysis.—Means and one standard deviation about the mean were arranged in Dice-Leraas diagrams to determine whether or not clinal variation in size existed in a north-south or east-west direction. The 44 populations shown in Figure 1 were grouped by species and used in the analysis. Characters evaluated were GLS, MT, ZB, BB, and FA.

Of the species recognized in this study, *rufus* and *sinaloae* had the most extensive distributions, but clear north-south clines were not evident in either. However, populations of *rufus* (1) and *sinaloae* (26) from northwestern México

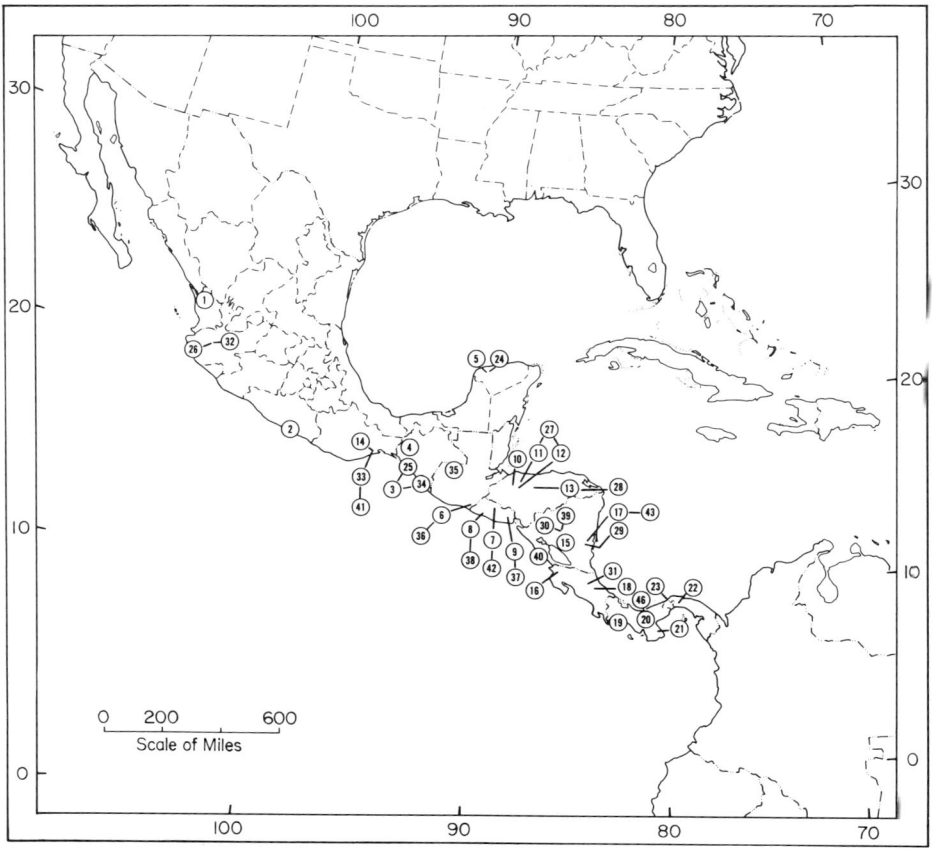

Fig. 1—Geographic locations of the 44 Central American populations for which electro-phoretic and morphological data were gathered. See text for a precise listing of localities.

were noticeably larger for all characters when compared to nearest geographic neighbors, as though populations in that region were isolated. Furthermore, a rough east-west difference was noted for *rufus*: populations (5, 10–13) from the Caribbean versant averaged large for the species, and it was to this size group that population 1 was most closely related, despite the fact that it lies west of the Sierra Madre Occidental.

Although population 17 from Nicaragua averaged larger than 18 from Costa Rica in every character plotted for *M. bondae*, the examination of only two populations precludes generalizing this observation to a clinal trend. Interestingly enough, the closely related species *pretiosus* showed the same tendency for the more northern locality to be largest, but the association was not as strong—for MT and BB, in males, Nicaraguan specimens were actually smaller.

An apparent cline for male *M. coibensis* was revealed in Panamá where populations 19 and 20 to the west exceeded 21–23 in overall size for all characters except FA. Females did not show as obvious a trend.

For *M. aztecus*, specimens from Nicaragua (39) were consistently larger than those from México (32), and bats from Guatemala (35) were either intermediate or of size equal to those from Nicaragua. Extremely small sample sizes should make the reader cautious in interpreting these data as an indication of an actual cline.

A mosaic pattern of geographic variability is seen in *M. molossus*. The only discernible pattern among populations was that those from southern México (33–35, 41) were relatively large, whereas Middle American populations in Guatemala (36) and El Salvador (37–38) were smaller. Specimens from Nicaragua (40) showed a return toward larger size.

Multivariate analysis.—Morphological variation among populations (= localities) was tested with a MANOVA, and populational relationships were described by means of a canonical variates analysis. The first survey was restricted to those populations sampled from Central America (Fig. 1), which, partitioned by sex, yielded 237 males representing 39 localities and 341 females from 42 localities All 16 variables listed under materials and methods were included. Inter- and intraspecific relationships presented in a bivariate plot of populational centroids along the first two canonical variates were mirrored in a more extensive morphological analysis outlined below. Because results did not differ between the two, only the latter analysis is included and discussed.

One objective of any taxonomic review necessarily must be a clarification of nomenclatorial uncertainties where possible. Of the plethora of specific epithets available within the genus *Molossus*, most apply to specimens from some place other than Central America, that is, the type localities occur principally outside the geographic region encompassed by this study. Ascertaining which names should be applied to the morpho-groupings identified in Central America thus necessitated broadening the original data base (618 total specimens) to incorporate populations from South America and the Greater and Lesser Antillean islands. Additionally, measurements for 16 type specimens were collected (D. C. Carter, unpublished data; Carter and Dolan, 1978), each type being treated as a new, unique operational taxonomic unit (OTU) (population). To include as many types as possible in the MANOVA, character states considered were reduced from 16 to the 10 listed in Tables 5 and 6, but the deletion of variables in no way altered earlier findings in terms of relationships of species, geographic variation, or character loadings.

All four MANOVA test criteria (Hotelling-Lawley Trace, Pillai's Trace, Wilke's Criterion, Roy's Maximum Root Criterion) overwhelmingly favored rejection of the null hypothesis with P less than or equal to 0.0001, indicating significant morphological differences among localities. This was true for both males and females.

TABLE 5.—*Normalized vector coefficients (eigenvalues) of canonical variates I and II showing the percentage influence of each variable in a MANOVA examining differences between localities for all populations of male* Molossus, *including type specimens.*

Character	Median	Vector I		Vector II	
		Eigenvalue	Percent influence	Eigenvalue	Percent influence
GLS	20.1	0.0518	17.61	−0.0779	16.57
CB	17.9	0.0222	6.72	−0.0198	3.75
BB	9.8	0.0807	13.38	−0.0867	8.99
PC	4.0	−0.0032	0.22	−0.2123	8.98
MT	7.0	0.1560	18.48	0.1796	13.30
BM	8.7	0.0501	7.38	−0.1322	12.17
BC	5.2	−0.0559	4.92	−0.0006	0.03
FA	43.7	0.0175	12.94	0.0487	22.52
MET3	45.2	0.0169	12.92	0.0223	10.67
MET4	43.8	−0.0073	5.41	−0.0065	3.02

TABLE 6.—*Normalized vector coefficients (eigenvalues) of canonical variates I and II showing the percentage influence of each variable in a MANOVA examining differences between localities for all populations of female* Molossus, *including type specimens.*

Character	Median	Vector I		Vector II	
		Eigenvalue	Percent influence	Eigenvalue	Percent influence
GLS	18.9	0.0633	18.92	−0.0708	18.23
CB	16.9	−0.0003	0.08	−0.0252	5.80
BB	9.5	0.0480	7.22	−0.0389	5.04
PC	3.9	−0.0005	0.03	−0.1266	6.73
MT	6.7	0.0552	5.85	0.1861	16.99
BM	8.3	0.0044	0.59	−0.1314	14.86
BC	4.8	0.0510	3.88	−0.0443	2.90
FA	42.4	0.0424	28.45	0.0266	15.37
MET3	44.0	−0.0282	19.64	0.0177	10.61
MET4	42.7	0.0228	15.41	0.0060	3.49

Ten canonical variates (characteristic roots) were extracted from the variance–covariance matrix for the 10 variables and 39 (males) or 42 (females) populations examined. The first two canonical variates expressed 94.9 percent and 95.7 percent of the total phenetic variation in males and females, respectively. Shown in Figures 2 and 3 are the population means plus one standard deviation; single specimens are denoted by solid dots, types by half circles. Dotted lines were used for clarity in areas of broadly overlapping populations. Seven distinct groups are evident, and the clusters are taken here as representing the following species of *Molossus*: *molossus* (A, populations 33–34, 36–38, 41–42, 44, 46, 51–54, 57, 59–62, 64–66), *sinaloae* (B, 25–31, 43), *coibensis* (C, 19–23, 50, 56), *aztecus* (D, 32, 35, 39), *bondae* (E, 17–18), *pretiosus* (F, 15–16), *rufus* (G, 1–14). Vector I, which accounts for the greatest percentage of total variation and is a measure of overall size, differentiates

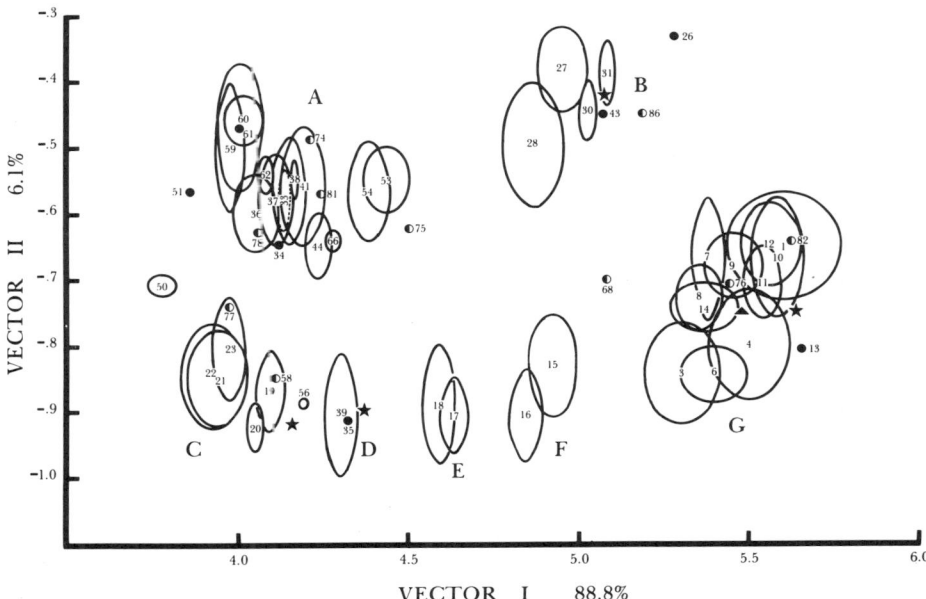

Fig. 2.—Projection of the first two canonical variates for 63 populations of male *Molossus*. Populational designations correspond to the position of mean values; ellipses represent one standard deviation about the mean. Seven species groupings are recognized and labelled A–G; A, *M. molossus;* B, *M. sinaloae;* C, *M. coibensis;* D, *M. aztecus;* E, *M. bondae;* F, *M. pretiosus;* G, *M. rufus.* Solid dots refer to single specimens, whereas half circles denote types. Population 63 is superimposed on 62 and hence not shown. Dotted lines were used in areas of broadly overlapping populations for clarity. Stars and triangles denote collections made between 1963 and 1967. See Table 1 for a precise listing of localities.

group A from B and groups C through G from one another. Populations forming the A and B clusters were separated from the remaining recognizable groups along Vector II, a shape-related component.

Among males, *M. rufus* is the largest taxon followed closely, in decreasing order, by *M. pretiosus*, *M. bondae*, *M. aztecus*, and *M. coibensis*. The importance of size in species discrimination is evident in the percent influence exerted by such size-related variables as MT, GLS, BB, and FA shown in Table 5. *M. sinaloae* and *M. molossus* separate along the second canonical variate axis primarily because they have a narrower skull. This difference in skull breadth between the A–B and C–G groups, depicted in Figure 2, is reflected in a marked increase in the contribution of breadth across molars (BM) in defining Vector II (see Table 5). Employing a size-out (independent) analysis of distance values, Freeman (1981) detected a similar phenetic clustering of *sinaloae* and *molossus*. Although Freeman did not elaborate on the origin of this pattern, unpublished information in hand suggests to me that it is correlated with a difference in food habits—*sinaloae* and *molossus* both preferring soft-bodied to hard-bodied flying insects.

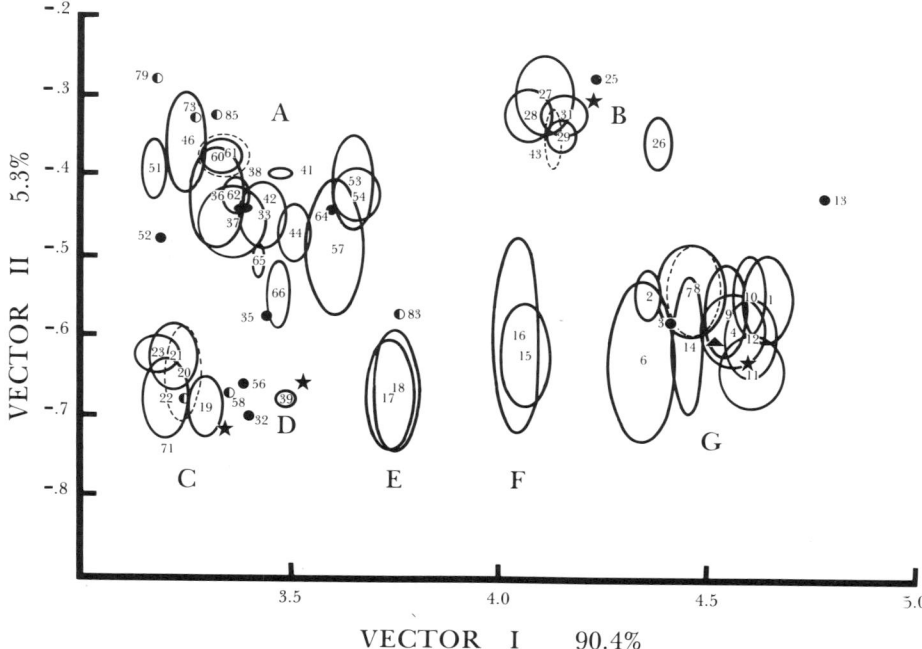

Fig. 3.—Projection of the first two canonical variates for 65 populations of female *Molossus*. See Figure 2 for an explanation of symbols.

Females, although showing the same specific relationships as males, differed slightly in the character loadings for Vector I. Whereas MT contributed heavily to the first canonical variate in males, in females it was of little importance (Table 6) and instead emphasis was shifted to wing elements (FA, MET3, MET4). Characters significant in describing Vector II, especially with respect to BM, were the same for both sexes.

Synonymies were generated by noting with which major cluster an individual type was associated. In most instances, there was little question as to which morphotype was represented. However, in the case of *M. pretiosus*, the holotype was intermediate between Central American populations of the species and *M. rufus* (see Fig. 2). Five paratypes (see the species account for *pretiosus*) of this taxon were included in a separate MANOVA analysis to determine if, by chance, the type for *pretiosus* could be considered a synonym of *rufus*. The answer was negative—paratypes of *pretiosus* from Venezuela more closely resembled populations from Central America referred to that species than they did populations of *rufus*. More detailed discussions of synonymies, where necessary, are dealt with under the appropriate species account.

Interesting patterns of geographic variation emerged from Figures 2 and 3. For *M. rufus*, for example, population 1, which is large and from northwestern México, was most closely associated with 9–12 (Honduras) and not with its nearest geographic neighbors, 3 and 4. Populations (7–8 El Salvador, 10–12 Honduras, 2–3 México) in proximity were quite similar in their morphology, but small increases in distance between localities sampled quickly resulted in discrete, essentially nonoverlapping clusters (compare populations 4 and 14 to 2–3, 9 and 6 to 7–8, 13 to 10–12).

The same influence of distance on morphological similarity was apparent in *M. coibensis* and *M. sinaloae*. For male *coibensis* in Panamá, populations 19 and 20, nearest neighbors, most closely resembled each other and were mensurally distinct from the more distant populations 22 and 23. Honduranian specimens of *sinaloae* (27–28) were slightly smaller than those from Nicaragua (29–30) or Costa Rica (31) to the south and smaller than Mexican samples (25–26) to the north and west.

Central American populations of *M. molossus* (33–34, 36–38, 41–42) were relatively uniform in their morphology; the most distinctive locality was 44 in Nicaragua. When the data base was expanded to include localities outside the Middle American region, surprising results were obtained. Specimens from the Lesser Antillean islands and Trinidad (59–62) were slightly smaller than, but quite similar to, Central American *molossus*. However, geographically intermediate Venezuelan bats (51–52) were dramatically diminutive. On the other end of the spectrum were Peruvian and Ecuadoran populations (53–54, 57), which averaged quite large for the species. Populations from the Greater Antilles (64–66) were most like those from Central America but showed a tendency to be somewhat larger. In general, four subclusters of *M. molossus* appeared discernible in Group A of Figures 2 and 3—the extremely small Venezuelan bats, the somewhat larger specimens from the Lesser Antilles, the medium-sized populations of Middle America, and the large bats from the Greater Antilles and South American mainland.

M. pretiosus and *M. bondae* formed discrete species-related clusters, but with only two populations sampled for each, little can be said regarding geographic variation other than that females tended to exhibit less interlocality variation than did males.

Although plagued by small sample size, *M. aztecus* (Group D, 32, 35, 39) was recognizable both morphologically and genetically as a species. In hand, live specimens were most easily confused with *M. molossus*, but as Figures 2 and 3 show, morphologically the taxon is more closely allied with *M. coibensis*. When a canonical variates analysis was performed on Central American populations only, females of population 35 from Guatemala clearly grouped with 32 and 39; its association with Jamaican specimens in Figure 3 is deemed a reflection of small sample size and the overall similarity among all species of small *Molossus*, and the consequent difficulty with which they are separated.

Karyology

Standard, G-, and C-banded karyotypes were obtained for the following species of *Molossus;* the point of origin of each is given in parentheses: *rufus*

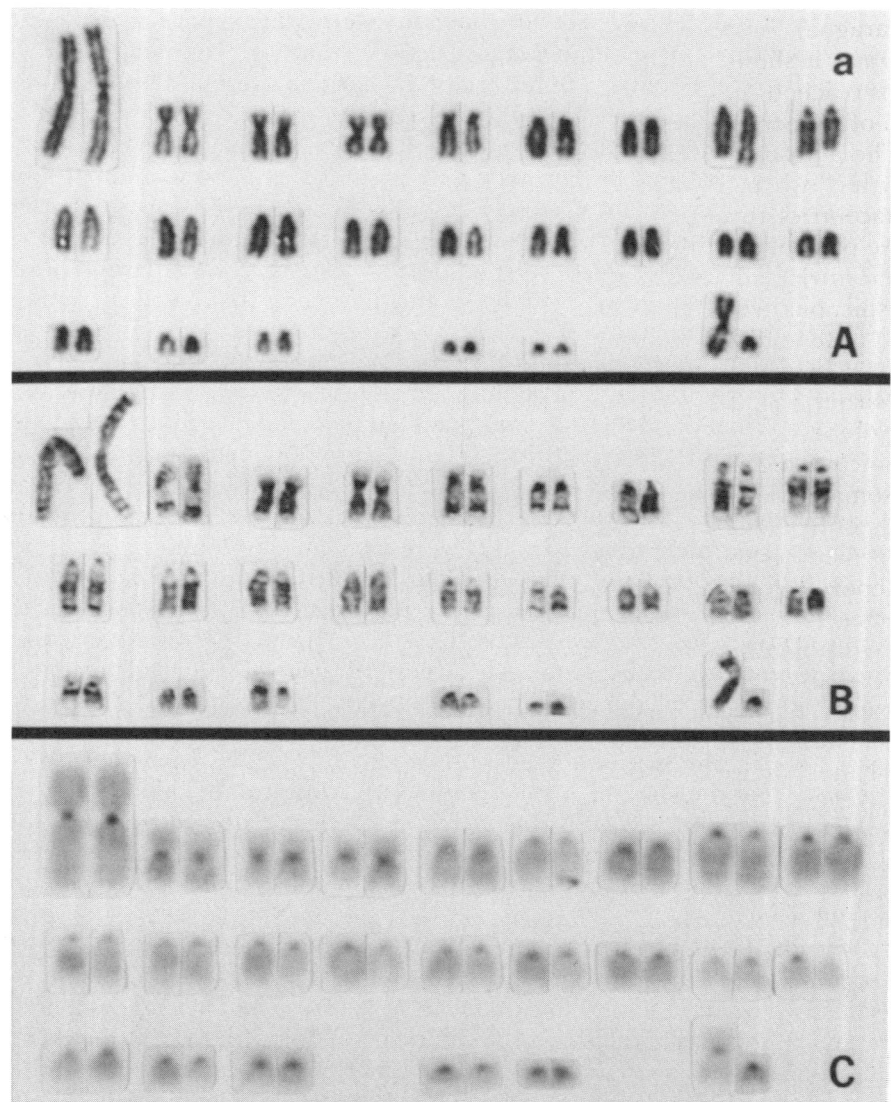

FIG. 4.—Standard (A), G-banded (B), and C-banded (C) karyotype of a male *Molossus rufus* from Chiapas, México (TTU 29472). Chromosomal morphology and banding patterns are considered representative of the genus.

(6 km. E Cintalapa de Figueroa, Chiapas, México), ¯*sinaloae* (Rama, Zelaya, Nicaragua), *molossus* (6 km. NE Rivas, Rivas, Nicaragua). The karyotype for *M. rufus* is shown in Figure 4 as representative of all three taxa, but conclusions reached regarding chromosome morphology are based on an examination of numerous spreads from all three species.

The autosomal complement of *M. rufus* contains one large pair of metacentrics, two pairs of medium-sized submetacentrics, one pair of medium metacentrics, three pairs of medium-sized subtelocentrics, 11 telocentric pairs, three pairs of small subtelocentrics (although one pair almost qualifies as submetacentric), and two pairs of small telocentrics. Secondary constrictions are conspicuous in one pair of large telocentric chromosomes. The diploid number (2N) is 48 and the fundamental number (FN), 66. The X is a medium-sized submetacentric and the Y is a small telocentric, or possibly subtelocentric.

Warner *et al.* (1974) were unable to detect differences between the standard karyotypes for species of *Molossus* they examined, and I also was unsuccessful in uncovering inter- or intraspecific variation. My interpretation of chromosomal morphology differs from that presented by the aforementioned authors only in the number of pairs considered subtelocentric, which was six. This increases the FN from 56 and 58 to 66 for all species examined here.

Reported for the first time are G- and C-bands for a member of the genus *Molossus* (Fig. 4, B–C). Most of the chromosomes in *Molossus* are large and show distinctive G-banding patterns that should prove useful in identifying the course of chromosomal change within the family. The telocentric chromosomes bearing secondary constrictions are thought to be represented by the G-banded pair labelled "a" in Figure 4.

C-bands showed heterochromatin was entirely centromeric (although a single spread of *M. molossus* suggested that the secondary constrictions stained positive). All subtelocentric short arms proved to be euchromatic. No differences in either G- or C-band patterns were evident among the three species examined.

Genetic Profile

Genetic similarity within and among species.—Data summarizing electrophoretic similarity among populations are shown in Figure 5 as a dendogram of Rogers' *S* values. Populations within a species are similar in allelic composition with *S* readings on the order of 95 percent or greater in some cases (as for *M. pretiosus, M. coibensis,* and *M. aztecus*). Figure 5 also indicates the existence of several distinctive species clusters, the most notable being that of *M. rufus, M. pretiosus,* and *M. bondae,* hereafter referred to as the *rufus* complex. In fact, no recognizable genetic difference exists between *pretiosus* and *rufus,* and *bondae* has but a single species-specific marker allele, this at the LDH locus. Additionally, specimens from locality 14, presumed topotypes of *M. pretiosus macdougalli,* are genetically indistinguishable from *M. rufus. M. coibensis* and *M. aztecus* possess a high genic similarity primarily due to a common PGM-1

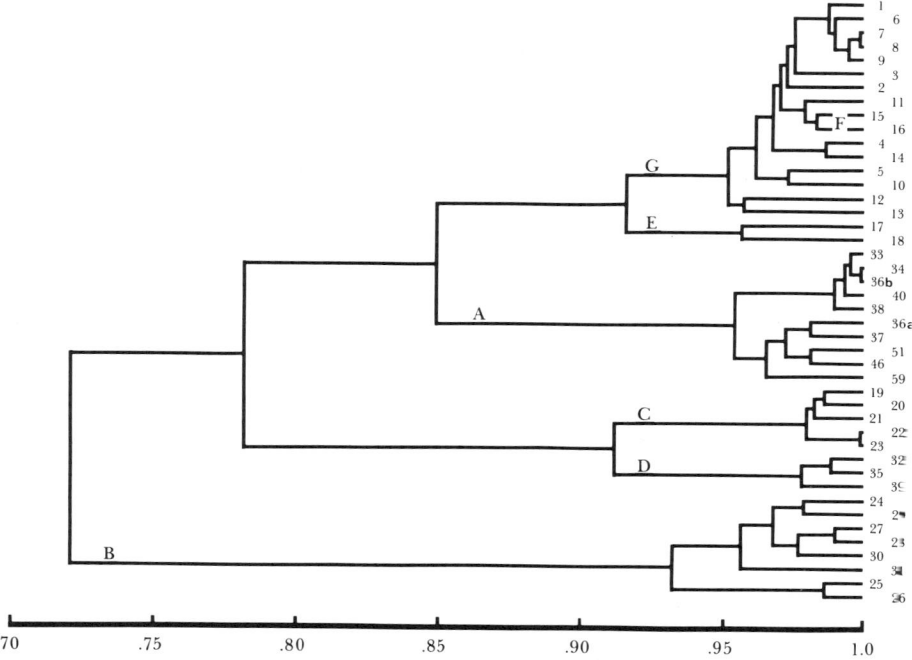

Fig. 5.—Phenogram of electrophoretic similarity values (Rogers' *S*) for Middle American populations of *Molossus*. See text for a key to localities. Letters designate species groupings: A, *M. molossus;* B, *M. sinaloae;* C, *M. coibensis;* D, *M. aztecus;* E, *M. bondae;* F, *M. pretiosus;* G, *M. rufus.*

and IPO-3 allele, allozymes shared also with *M. sinaloae*. However, genetic identity is maintained because each bears a unique allele or allelic combination shown in parentheses: *M. coibensis* LDH locus, (98); *M. aztecus* EST-2 locus (98). *M. sinaloae* is the most divergent taxon because of unique isozymes at the EST-2, α-GPD, and MDH loci.

There is not a strong correlation between genetic similarity and geographic proximity (Fig. 5). For example, population 1 of *M. rufus* groups with material from Guatemala (6) and El Salvador (7–9) before it is associated with specimens from México (specifically populations 2, 4, 14); population 11 is more nearly like *M. pretiosus* than its own conspecifics in close proximity (populations 10, 12–13). The same argument can be presented for *M. aztecus* where Guatemalan bats (population 35) are more like those from México (32) than those from nearby Nicaragua (39). Yucatanian *M. sinaloae* (24) more closely resemble specimens from Rama, Nicaragua (29) than geographically intermediate material from Honduras (27–28) or northern Nicaragua (30).

The failure of distance to track genetic similarity suggests populations are isolated. This is supported by the occurrence of rare alleles and marked variation in gene frequency data (Table 7). By way of illustration, alleles known to be present in only one population have been recorded for the following species (locus is followed by the specific allele designation and population in which it exists): *M. rufus* (GP, 122, 11; EST-2, 132, 11; EST-2, 143, 11); *M. pretiosus* (PGM-1, 108, 15); *M. coibensis* (PGM-1, 97, 23); *M. sinaloae* (EST-2, 117, 24; GP, 109, 31; PGM-1, 103, 27); *M. molossus* (α-GPD, 136, 51; EST-2, 104, 51). Other isozymes are localized in occurrence, such as IDH 127 in populations 10 and 11 of *M. rufus* and PGM-1 89 in populations 24, 26, and 31 of *M. sinaloae*. Intraspecific variation in allelic frequency is demonstrated in *M. bondae* at the PGM-1 locus, allele 85, which exists at a six percent level in population 17, and a 40 percent level in population 18. The 116 allele in *M. coibensis* accounts for 43 percent of the EST-2 locus composition in population 20 compared to zero to 21 percent in the remaining four populations. In *M. sinaloae*, the α-GPD 106 allele is virtually fixed in population 31 but absent in populations 25 and 26. Individuals of Nicaraguan (40) *M. molossus* exhibit the EST-2 127 at a frequency of 98 percent, but its occurrence drops dramatically to 18 percent in neighboring El Salvador (37). Additional examples are given in Table 7.

Isoelectric focusing.—Genetic similarity can be overestimated by starch-gel electrophoresis because the technique operates by separating proteins on the bases of charge and size, variables that can offset one another and lead to equivalent migration rates. A more sensitive assay for allelic congruence, isoelectric focusing, was applied to the esterase locus to demonstrate the commonality of apparent genic synapomorphs and to utilize the refined resolving powers of electrofocusing for more fully characterizing an important species-discriminating isozyme. Verification of interspecific allelic identity by independent tests, when possible, heightens the accuracy of phylogenetic inferences drawn from genic data. Although several loci were tested initially, due to the nature of the difficulties in working with native gels (gels in which enzymatic activity is preserved) reliable and consistent results were obtained only for esterase.

A total of four esterase gels were run with samples from all species, save *M. aztecus*, emphasizing four major alleles—100, 108, 116, 131. Interspecific comparisons showed the 100, 116, and 131 alleles were identical in *M. rufus*, *M. pretiosus*, *M. bondae*, *M. molossus*, and *M. coibensis*. The 108 isozyme, species specific for *M. sinaloae*, was included as a reference point.

A composite from several gels (Fig. 6) shows densitometer readings for three molossid species, *M. molossus*, *M. rufus*, and *M. sinaloae*. Peaks represent focused bands on a vertical slab polyacrylamide medium; height is proportional to the concentration of functional enzyme present. The tracings overlay perfectly even though the samples differ in electrophoretic designation: *molossus* 116; *rufus* 116, 131; *sinaloae* 108. At the esterase locus, species apparently differ not so much in the presence or absence of functional alleles but in the relative activity of a broad spectrum of isozymes, which points to specific differentiation in the regulatory mechanisms governing gene expression or the importance of post-transcriptional modification.

TABLE 7.—*Alleles and frequencies (given in parentheses) for 11 polymorphic loci in seven species of Molossus. Monomorphic loci included EST-2, –SDH, Alb, Hb, GOT, –GOT, PGM-3, G6P, GDH. See Table 1 for a key to localities and text for a description of locus designations.*

M. rufus

N	Loc	EST-2	α-GPD	–ADH	IDH	MDH	–MDH	GP	LDH	IPO2	PGM1	–IPO3
21	1	100(0.86), 148(0.14)	100(1.00)	–100(1.00)	100(1.00)	100(1.00)	–100(1.00)	100(1.00)	100(1.00)	100(1.00)	103(0.93), 108(0.07)	–100(1.00)
5	2	100(0.80), 132(0.20)	100(0.80), 122(0.20), 100(0.67)	–100(1.00)	100(1.00)	100(1.00)	–100(1.00)	100(1.00)	100(1.00)	100(1.00)	103(0.80), 108(0.20)	–100(1.00)
3	3a	100(1.00)	122(0.33)	–100(1.00)	100(1.00)	100(1.00)	–100(1.00)	100(1.00)	100(1.00)	100(1.00)	103(1.00)	–100(1.00)
7	3b	86(0.05), 100(0.90), 131(0.05)	100(0.90), 122(0.10)	–100(1.00)	100(1.00)	100(0.95), 120(0.05)	–100(1.00)	100(1.00)	100(1.00)	100(1.00)	103(1.00)	–100(1.00)
20	4	86(0.03), 100(0.45), 131(0.52)	100(0.90), 129(0.10)	–100(1.00)	100(1.00)	100(1.00)	–100(1.00)	100(1.00)	100(1.00)	100(1.00)	103(0.88), 108(0.12)	–100(1.00)
10	5	100(0.56), 131(0.44)	100(0.50), 112(0.50)	–100(1.00)	100(1.00)	100(1.00)	–100(1.00)	100(1.00)	100(1.00)	100(1.00)	103(0.94), 108(0.06)	–100(1.00)
9	6a	100(0.89), 131(0.11)	100(0.89), 129(0.11)	–100(1.00)	100(1.00)	100(1.00)	–100(1.00)	100(1.00)	100(1.00)	100(1.00)	103(1.00)	–100(1.00)
1	6b	100(0.50), 131(0.50)	100(1.00)	–100(1.00)	100(1.00)	100(1.00)	–100(1.00)	100(1.00)	100(1.00)	100(1.00)	103(1.00)	–100(1.00)
23	7	100(0.86), 131(0.14)	100(1.00)	–100(1.00)	100(1.00)	100(1.00)	–100(1.00)	100(1.00)	100(1.00)	100(1.00)	103(1.00)	–100(1.00)
31	8	100(0.82), 131(0.18)	100(0.98), 112(0.02)	–100(1.00)	100(1.00)	100(1.00)	–100(1.00)	100(1.00)	100(1.00)	100(1.00)	103(1.00)	–100(1.00)
21	9	100(0.71), 116(0.02), 131(0.27)	100(1.00)	–100(1.00)	100(1.00)	100(1.00)	–100(1.00)	100(1.00)	100(1.00)	100(1.00)	103(1.00)	–100(1.00)
20	10	100(0.77), 116(0.05), 131(0.18)	100(0.58), 112(0.42)	–100(1.00)	100(0.83), 127(0.17)	100(1.00)	–100(1.00)	100(1.00)	100(1.00)	100(1.00)	85(0.03), 103(0.97)	–100(1.00)
19	11	100(0.67), 116(0.12), 131(0.13), 132(0.08)	100(0.84), 112(0.16)	–100(1.00)	100(0.92), 127(0.08)	100(0.97), 120(0.03)	–100(1.00)	100(0.87), 122(0.13)	100(1.00)	100(1.00)	85(0.13), 103(0.87)	–100(1.00)
7	12	100(0.43), 116(0.14), 131(0.43)	100(0.71), 122(0.29)	–100(1.00)	100(1.00)	100(1.00)	–100(1.00)	100(1.00)	100(1.00)	100(1.00)	85(0.50), 103(0.50)	–100(1.00)
2	13	100(0.75), 143(0.25)	100(0.25), 112(0.75)	–100(1.00)	100(1.00)	100(1.00)	–100(1.00)	100(1.00)	100(1.00)	100(1.00)	85(0.50), 103(0.50)	–100(1.00)
7	14	100(0.29), 131(0.57), 135(0.14)	100(0.93), 122(0.07)	–100(1.00)	100(1.00)	100(1.00)	–100(1.00)	100(1.00)	100(1.00)	100(1.00)	103(0.86), 108(0.14)	–100(1.00)

TABLE 7.—*Continued.*

N	Loc	EST-2	α-GPD	-ADH	IDH	MDH	-MDH	GP	LDH	IPO2	PGM1	-IPO3
						M. pretiosus						
22	15	100(0.55) 116(0.15) 131(0.30)	100(0.86) 112(0.14)	-100(1.00)	100(1.00)	100(1.00)	-100(1.00)	100(1.00)	100(1.00)	100(1.00)	85(0.09) 103(0.91)	-100(1.00)
16	16	100(0.68) 116(0.04) 131(0.17) 132(0.04) 141(0.07)	100(0.84) 112(0.16)	-100(1.00)	100(1.00)	100(1.00)	-100(1.00)		100(1.00)	100(1.00)	85(0.13) 103(0.78) 108(0.09)	-100(1.00)
1	67	100(0.50) 116(0.50)	100(1.00)					100(1.00)		100(1.00)	103(1.00)	-100(1.00)
						M. bondae						
16	17	100(0.24) 116(0.08) 131(0.68)	100(0.88) 112(0.13)	-100(1.00)	100(1.00)	100(0.94) 108(0.06)	-100(1.00)	100(1.00)	88(0.03) 102(0.97)	100(1.00)	85(0.06) 103(0.94)	-100(1.00)
20	18	100(0.40) 116(0.30) 131(0.30)	100(0.90) 112(0.13)	-100(1.00)	100(1.00)	100(1.00)	-100(1.00)	100(1.00)	88(0.28) 102(0.72)	100(1.00)	85(0.40) 103(0.60)	-100(1.00)
						M. coibensis						
26	19	100(0.94) 116(0.06)	100(1.00)	-100(1.00)	100(1.00)	100(1.00)	-100(1.00)	100(1.00)	98(1.00)	-100(1.00)	89(1.00)	-167(1.00)
21	20	100(0.57) 116(0.43)	100(1.00)	-100(1.00)	100(1.00)	100(1.00)	-100(1.00)	100(1.00)	98(1.00)	-100(1.00)	89(1.00)	-167(1.00)
14	21	100(0.79) 116(0.21)	100(1.00)	-100(1.00)	100(1.00)	100(1.00)	-100(1.00)	100(1.00)	98(1.00)	-100(1.00)	89(1.00)	-167(1.00)
23	22	100(0.98) 116(0.02)	100(1.00)	-100(1.00)	100(1.00)	100(1.00)	-100(1.00)	100(1.00)	98(1.00)	-100(1.00)	89(1.00)	-167(1.00)
17	23	100(1.00)	100(1.00)	-100(1.00)	100(1.00)	100(1.00)	-100(1.00)	100(1.00)	98(1.00)	-100(1.00)	89(0.97) 97(0.03)	-167(1.00)

TABLE 7.—Continued.

M. sinaloae

N	Loc	EST-2	α-GPD	−ADH	IDH	MDH	−MDH	GP	LDH	IPO2	PGM1	−IPO3
13	24	108(0.83) 117(0.17)	79(0.33) 106(0.67)	−100(0.58) −113(0.42)	100(1.00)	100(1.00)	−73(1.00)	100(1.00)	100(1.00)	−100(1.00)	79(0.29) 89(0.04) 100(0.67)	−167(1.00)
1	25	108(1.00)	79(1.00)	−100(1.00)	100(1.00)	100(1.00)	−73(1.00)	100(1.00)	100(1.00)	−100(1.00)	79(1.00)	−167(1.00)
3	26	108(1.00)	79(1.00)	−100(1.00)	100(1.00)	100(1.00)	−73(1.00)	100(1.00)	100(1.00)	−100(1.00)	79(0.67) 89(0.17) 100(0.16)	−167(1.00)
12	27[a]	108(1.00)	79(0.54) 106(0.46)	−100(0.67) −113(0.33)	100(1.00)	100(1.00)	−73(1.00)	100(1.00)	100(1.00)	−100(1.00)	79(0.25) 100(0.67) 103(0.08)	−167(1.00)
1	27[b]	108(1.00)	79(1.00)	−100(1.00)	100(1.00)	100(1.00)	−73(1.00)	100(1.00)	100(1.00)	−100(1.00)	79(0.50) 100(0.50)	−167(1.00)
22	28	108(1.00)	79(0.52) 106(0.48)	−100(0.77) −113(0.23)	100(1.00)	100(1.00)	−73(1.00)	100(1.00)	100(1.00)	−100(1.00)	79(0.36) 100(0.64)	−167(1.00)
22	29	108(1.00)	79(0.40) 106(0.60)	−100(0.45) −113(0.55)	100(1.00)	100(1.00)	−73(1.00)	100(1.00)	100(1.00)	−100(1.00)	79(0.25) 100(0.75)	−167(1.00)
2	30	108(1.00)	79(0.50) 106(0.50)	−100(1.00)	100(1.00)	100(1.00)	−73(1.00)	100(1.00)	100(1.00)	−100(1.00)	79(0.50) 100(0.50)	−167(1.00)
25	31	108(1.00)	73(0.12) 106(0.88)	−100(0.80) −113(0.20)	100(1.00)	100(1.00)	−73(1.00)	100(0.96) 109(0.04)	100(1.00)	−100(1.00)	79(0.58) 89(0.12) 100(0.30)	−167(1.00)

TABLE 7.—*Continued.*

N	Loc	EST-2	α-GPD	-ADH	IDH	MDH	-MDH	GP	LDH	IPO2	PGM1	-IPO3
						M. aztecus						
1	32	98(1.00)	100(1.00)	-100(1.00)	100(1.00)	100(1.00)	-100(1.00)	100(1.00)	100(1.00)	-100(1.00)	89(1.00)	-167(1.00)
2	35	98(0.75) 127(0.25)	100(1.00)	-100(1.00)	100(1.00)	100(1.00)	-100(1.00)	100(1.00)	100(1.00)	-100(1.00)	89(1.00)	-167(1.00)
3	39	98(0.50) 116(0.50)	100(1.00)	-100(1.00)	100(1.00)	100(1.00)	-100(1.00)	100(1.00)	100(1.00)	-100(1.00)	89(1.00)	-167(1.00)
						M. molossus						
14	33*	116(0.11) 127(0.89)	112(1.00)	-100(1.00)	100(1.00)	100(1.00)	-100(1.00)	100(1.00)	100(1.00)	100(1.00)	103(1.00)	0(1.00)
2	34	127(1.00)	112(1.00)	-100(1.00)	100(1.00)	100(1.00)	-100(1.00)	100(1.00)	100(1.00)	100(1.00)	103(1.00)	0(1.00)
7	36ᵃ	116(0.50) 127(0.50)	112(1.00)	100(1.00)	100(1.00)	100(1.00)	-100(1.00)	100(1.00)	100(1.00)	100(1.00)	103(1.00)	0(1.00)
1	36ᵇ	127(1.00)	112(1.00)	-100(1.00)	100(1.00)	100(1.00)	-100(1.00)	100(1.00)	100(1.00)	100(1.00)	103(1.00)	0(1.00)
11	37	116(0.82) 127(0.18)	112(1.00)	-100(0.95) 113(0.05)	100(1.00)	100(1.00)	-100(1.00)	100(1.00)	100(1.00)	100(1.00)	103(1.00)	0(1.00)
4	38	116(0.25) 127(0.75)	112(1.00)	-100(1.00)	100(1.00)	100(1.00)	-100(1.00)	100(1.00)	100(1.00)	100(1.00)	103(1.00)	0(1.00)
24	40	116(0.02) 127(0.98)	94(0.08) 112(0.92)	-100(1.00)	100(1.00)	100(1.00)	-100(1.00)	100(1.00)	100(1.00)	100(1.00)	103(1.00)	0(1.00)
2	42	116(0.25) 127(0.75)	112(1.00)	-100(1.00)	100(1.00)	100(1.00)	-100(1.00)	100(1.00)	100(1.00)	100(1.00)	103(1.00)	0(1.00)
2	46	116(1.00)	94(0.25) 112(0.75)	-100(1.00)	100(1.00)	100(1.00)	-100(1.00)	100(1.00)	100(1.00)	100(1.00)	103(1.00)	0(1.00)
25	51	104(0.02) 116(0.87) 131(0.11)	94(0.09) 112(0.87) 136(0.04)	-100(0.98) -125(0.02)	100(1.00)	100(1.00)	-100(1.00)	100(1.00)	100(1.00)	100(1.00)	—	0(1.00)
21	59	100(0.10) 116(0.90)	112(1.00)	-100(1.00)	100(1.00)	100(1.00)	-100(1.00)	100(1.00)	92(0.38) 100(0.62)	100(1.00)	103(1.00)	0(1.00)

ᵃˑᵇ In a few instances, two populations were grouped to increase sample size for the morphological comparisons. The superscript "a" refers to the first locality in entries shown in Table 1 with multiple listings, and "b" refers to the second.

*This is a composite of populations 33 and 41 of Table 1.

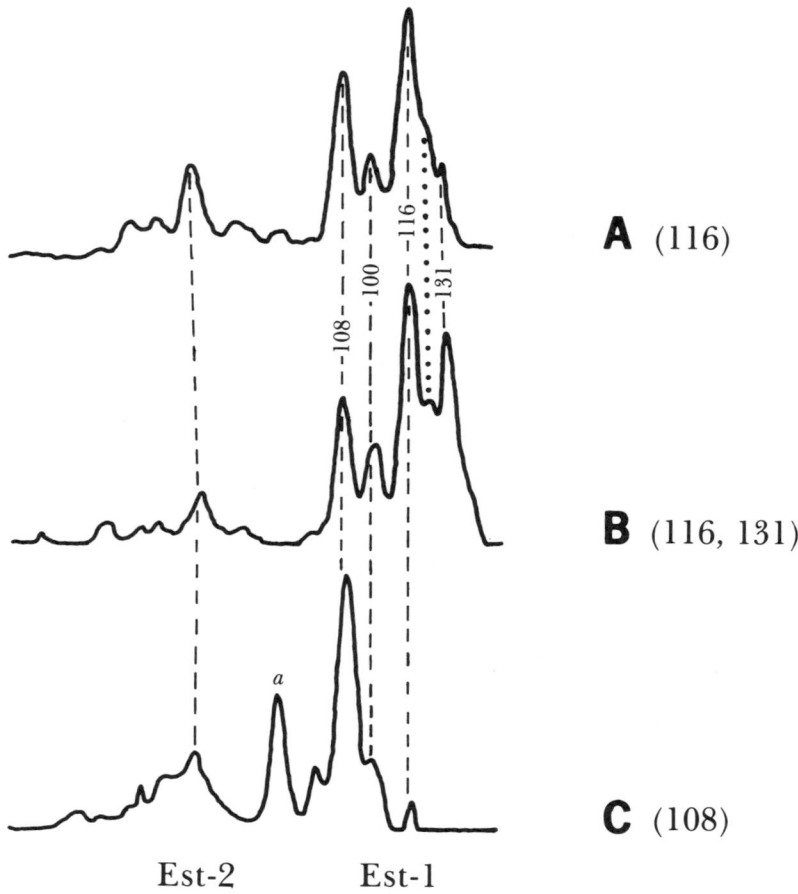

Fig. 6.—Selected densitometer tracings of the EST-1 and EST-2 loci from equilibrium poly-acrylamide gels for three species of *Molossus*: A, *M. molossus;* B, *M. rufus;* C, *M. sinaloae.* The corresponding EST-1 electrophoretic designation of alleles for each sample shown appears in parentheses. Dashed lines join homologous isozymes and are labelled according to the electrophoretic allele each represents; dotted lines join minor alleles resolvable only with isoelectric focusing. A unique minor allele present only in *M. sinaloae* is denoted by an italic a.

The peaks in Figure 6 are labelled according to the homologous alleles discernible in starch-gel electrophoresis. Note that the EST-2 locus is a constant among taxa. The most intense EST-1 band within each taxon corresponds to the single allele resolved by electrophoretic techniques. However, *M. rufus,* illustrated by 6B is heterozygous (116, 131) and consequently shows two strong peaks. The concentration of the initial protein mixture for isoelectric separation as well as the linearity of the pH gradient established in the gel will affect the clarity of focused bands and may even obscure the resolution of minor bands (note the variation in intensity of minor bands in Fig. 6). Based on iso-

TABLE 8.—*Isoelectric pH values for several major esterase alleles of Middle American* Molossus.

Locus	Electrophoretic allele	Isoelectric pH
Est-2	100	6.25
Est-1	a*	6.00
	108	5.70
	100	5.65
	116	5.50
	131	5.40

* A unique minor allele in *M. sinaloae* not resolved by electrophoresis.

electric data, *M. sinaloae* appears unique among the molossids in two respects—the existence of an extremely active isozyme ("a" in Fig. 6) essentially lacking in the other taxa and the absence of any trace of the 131 allele.

Isoelectric pH values for the major esterase alleles are presented for both loci in Table 8 and are shown in the order of increasing negative charge. Like other globular proteins, the esterase alleles have slightly acidic isoelectric points, pointing to a preponderance of acidic residues in their amino acid composition. The ranking presented in Table 8 mimics the relative migration rates of the alleles seen on starch gel with one exception. In a starch support matrix, allele 108 exceeds the 100 allele in anodal movement, but their positions are transposed on a polyacrylamide slab. This suggests that although the 100 allele has a greater negative charge, it may be sufficiently larger in size or bulkier in its tertiary configuration so that its movement is retarded in a starch gel.

Genetic differentiation among populations.—Significant differentiation among populations for almost every species of *Molossus* was revealed by an analysis of the variance in gene frequencies (F_{ST}) among localities as shown in Table 9. Three loci, PGM-1, α-GPD, and EST-2, in particular contributed to the intraspecific variability. Both F_{IT} and F_{IS} can be regarded as inbreeding coefficients (Wright, 1978). Positive values for F_{IT} are related to systematic subdivision of the total populace (Wright, 1965). Only *M. bondae* showed a negative F_{IT}, which may be a consequence of sampling error, given that only two populations were represented. However, until behavioral studies are done, it is possible that a difference in social organization exists between *bondae* and other species of *Molossus* such that there is an avoidance of consanguinous matings within populations of *bondae*. For the species listed in Table 9, F_{IT} ranges from .0874 to .4846 and indicates a greater number of homozygous individuals than would be expected if data for all populations were pooled, suggesting that each species is composed of recognizably different subunits (= populations) on the basis of gene frequency variation. When F_{IS} is positive, an excess of homozygous individuals within the hierarchy under consideration is indicated reflecting either inbreeding or a further subdivision of that level. Loci with positive values appear to vary randomly among the species. With the ex-

TABLE 9.—*Genetic variation among populations within species of* Molossus *as revealed by F-statistics. Values represent means calculated over intraspecific polymorphic loci only and not over all generically variable loci. Exchange rates required to maintain the levels of populational differences indicated by F_{ST} are also given.*

Species (no. populations)	Locus	F_{IT}	F_{IS}	F_{ST}	Chi-square	df	Individuals per generation
M. rufus (14)	GP	.5085	.4559	.0966	80.8***	26	
	IDH	.0015	-.1415	.1253	52.4***	13	
	PGM-1	.1373	-.0360	.1673	419.6***	78	
	α-GPD	.4034	.1951	.2588	713.7***	91	
	EST-2	.3637	.2749	.1225	601.8***	169	
	TOTAL	.2829	.1497	.1541	1868.2***	377	1.4
M. pretiosus (2)	PGM-1	.0998	.0767	.0250	11.4*	6	
	α-GPD	.2738	.2732	.0008	.4	7	
	EST-2	.2483	.2266	.0281	26.3**	13	
	TOTAL	.2073	.1922	.0180	38.2*	26	13.6
M. bondae (2)	LDH	.0256	-.0894	.1056	30.4***	4	
	PGM-1	.1343	-.0185	.1500	64.8***	6	
	α-GPD	.0934	-.0951	.0016	.8	7	
	EST-2	-.1014	-.1380	.0322	20.9*	13	
	TOTAL	-.0349	-.0853	.0724	116.6***	30	3.2
M. coibensis (5)	PGM-1	.0246	0	.0246	29.8	24	
	EST-2	.2545	.0216	.2381	625.1***	52	
	TOTAL	.1396	.0108	.1314	654.9***	76	1.6
M. sinaloae (8)	PGM-1	.0903	-.0921	.1670	198.4***	42	
	α-GPD	.1144	-.0667	.1698	247.3***	49	
	EST-2	.0576	.0366	.0218	54.4	91	
	TOTAL	.0874	.0407	.1195	500.1***	168	1.8
M. aztecus (3)	EST-2	.4571	.2963	.2286	35.7*	26	.8
M. molossus (9)	LDH	.4754	.2115	.3347	307.9***	32	
	PGM-1	.8739	.1812	.8460	1116.8***	48	
	α-GPD	.0057	-.0574	.0596	93.5*	56	
	EST-2	.5834	-.1291	.6310	1886.9***	104	
	TOTAL	.4846	.0516	.4678	3404.9***	240	0.8

*$P < 0.10$; **$P < 0.025$; ***$P < 0.001$.

TABLE 10.—*Summary of average gene diversity attributable to various hierarchial levels: populations within species (G_{PS}), species within total (G_{ST}). Diversity was assessed first using all 11 variable loci (interspecifically variable) and then the analysis was repeated on a more restricted data set (intraspecifically variable) due to the occurrence of species-specific alleles.*

Polymorphic loci	G_{PS}	$\bar{G}_{ST}$
Interspecifically variable	0.1224	0.5342
Intraspecifically variable	0.1684	0.3596

ception of *M. bondae,* populations within species of *Molossus* demonstrate some degree of inbreeding, which is indicated by F_{IS} ranging from .01 to .29. Values for *M. aztecus* could be inflated due to extremely small sample size.

An analysis of gene diversity (Nei, 1973) within populations and ultimately within species is presented in Table 10. If all polymorphic loci are considered, then approximately 12 percent of the variation in allelic frequencies can be attributed to differences among populations (G_{PS} = .1224), which is to say that only 88 percent ($1 - G_{PS}$) of the total gene diversity is vested within any one population. Species account for most of the variation in gene frequencies (G_{ST} = 0.53), as would be expected due to the existence of a number of species-specific alleles. Species of *Molossus* show considerable genetic differentiation, and only 47 percent of the total variation of allelic composition ($1 - G_{ST}$) can be found in any single species. Half of the polymorphic loci are specific to only one taxon; therefore, a more meaningful measure of populational diversity within the genus would be one in which only intraspecifically variable loci were included. When this is done, differentiation among populations within a species increases to G_{PS} = 17 percent.

DISCUSSION

Species groupings recognized herein are based on a refined understanding of intraspecific variation (both morphological and genetic) and a conservative interpretation of this variation (see individual species accounts for summary remarks on systematic relationships).

Distribution

Few species of *Molossus*, as understood here, are continuously distributed. The appearance of species in isolated mountain ranges, on opposite versants between Central and South America, and in widely separated geographic regions is explained by Pleistocene climatic events and the impact of those changes on the expansion and contraction of vegetative zones, particularly woodlands.

Generic progenitors undoubtedly reached North America by early Miocene via the Bering Strait before the climate of that passageway cooled sufficiently to loose its tropical-warm temperate nature (Koopman, 1970). Orogenic activity begun in the late Eocene along the eastern margin of the Middle American Trench that stretched from what is now southern Nicaragua to northern Colombia culminated in earliest Pliocene in a continuous land connection that would have allowed molossids to move freely southwards. Where the genus *Molossus* originated, whether it was Middle America or South America, and when the taxon first became recognizable may never be known, but it is almost certain that faunal exchanges would have occurred unimpeded between the two regions from the Pliocene onward; the isolation and fractioning of populations induced by bioclimatic fluctuations during the Pleistocene, however, probably had more to do with the differentiation of species and the present day distributional patterns than any preceding set of geological events. Haffer (1967b) demonstrated the effect such glacial-interglacial interludes had on the distribution of nonforest lowland bird faunas in South America, and Duellman (1960, 1966) invoked the same principles to explain the distributions of reptiles and amphibians in Central America.

Species of *Molossus* can be divided almost evenly into forest (*bondae, aztecus, sinaloae*) and nonforest (*rufus, pretiosus, coibensis, molossus*) dwellers, nonforest habitats being broadly interpreted in the sense of Haffer (1967b) as open areas like grassland savannas, dry open woodlands, cactus wastes, thorn scrub, and so forth. The zoogeographic relations of mastiff bats within these two biotic assemblages was shaped by the ebb and flow of the nonforest biotypes. A general three to four degree centigrade lowering in average temperature helped drive these changes, but primarily they were the result of advances and retreats in the breadth of the equatorial rainbelt that occurred during glacial and interglacial periods, respectively (Haffer, 1967a). These alternations between dry and humid climates apparently continued into post-Pleistocene times.

Of the forest taxa, *M. bondae* closely follows tropical rainforest habitat along the Caribbean lowlands in Middle America as far as northern Honduras where the forest becomes restricted on the coast; the species is likely to be ubiquitous throughout the area. Its appearance in western and southwestern Colombia along the Pacific versant is due to the "crossing over" of the rainforest belt in Panamá, a phenomenon first described by Dunn (1940). The only other South American record for *bondae* is that from the type locality of Santa Marta on the northern Colombian coast. This population is probably a relict left by the last interglacial episode when forests contracted inasmuch as arid environments now surround the locality. Other populations may persist in refugia scattered about Colombia's intermontane valleys and in portions of western Venezuela, but it seems unlikely that this taxon ever occurred much farther east than Venezuela.

The Isthmus of Tehuantepec historically has been an effective barrier to species movement but it must have been traversed on occasion in order to explain herpetological patterns of distribution (Duellman, 1960, 1966). Climatic fluctuations during the Pleistocene that altered the vegetation within this pass alternately opened and closed corridors for dispersal. *M. aztecus* is a montane bat occurring both north and south of the Isthmus. Although now disjunct, its distribution was once probably continuous as a result of glacial periods that permitted the encroachment of upland pine-oak habitat into the isthmian region. So far, *M. aztecus* has not been taken in the mountains of Costa Rica or Panamá nor is it known with assurance from South America. It is quite possible that this taxon extends no farther south than Nicaragua, and that an interchange across the Nicaraguan lowlands never has taken place because of insufficient climatic depression of upland floras in this segment of the Central American cordillera.

Like *M. bondae*, *M. sinaloae* occupies rainforest habitats on the Caribbean side of the Middle American cordillera but it is found farther north and east. Populations on the Pacific coast in southern México and Nicaragua and in the vicinity of Puntarenas in Guanacaste, Costa Rica, probably arrived in these areas when Pleistocene forests stretched across the Isthmus of Tehuantepec and the Nicaraguan lowlands. Today they exist as isolates with intervening nonforest zones further subdividing their distribution. Farther north, dispersal was most likely east-west across México about the level of 19° latitude, with the taxon following a sweepstakes route through the mountains. In South America, *M. sinaloae* is found in mesic forest refugia in both the Cauca and Magdalena intermontane valleys of Colombia. I have no records for this species west of the Andean Cordillera Occidental, but its appearance in the Upper Patia Valley suggests to me that it inhabits that region, as does *bondae* with which it is sympatric in Central America, although it is possible it traversed the high plains of Popayan, like a number of birds (Haffer, 1967*b*), and actually came from the Cauca. During the more humid glacial stages, *sinaloae* was probably widespread along the northern coast of South America; however, in Recent times the advancement of nonforest biotypes has disrupted its distribution.

Nonforest mastiff bats appear more widespread than their forest counterparts because of the almost contiguous nature of the xeric habitat with which they are associated. *M. rufus* is equally represented on both versants in México and probably has enjoyed a comparatively free faunal exchange across the Isthmus of Tehuantepec in Recent times. The same is likely to be true of *M. molossus*, but its distribution is less well defined. Individuals of *rufus* could have reached the state of Sinaloa by either moving northward up the west coast from the Isthmus or could have come from the east by following the tortuous wanderings of the Balsas Basin. I deem the later avenue of dispersal more probable because of the better agreement in size between Sinaloan *rufus* and those from the Caribbean side and the absence of the α-GPD 122 allele, which is relatively common (10–33 percent) in populations directly to the south. The large *Molossus* occupying South America would seem to be conspecific with those inhabiting Middle America, but the link between the two regions that once must have existed has since been severed by the lowland rainforest that crosses over from the Caribbean to the Pacific versant in southern Panamá and northern Colombia. This area should be considered a filter zone for the entire assemblage of nonforest mastiff bats. North of Nicaragua, the distribution of *rufus* is fairly continuous along the west coast. From Nicaragua to Panamá it is fractured, apparently as a result of interactions with *M. pretiosus*, a species with which it must surely compete, and the intrusion of moist forest habitat especially in southern Costa Rica and western Panamá.

The most common mastiff bat in Panamá, *M. coibensis,* is poorly known outside of that country. I have allocated additional specimens to the taxon, including the types of *M. aztecus lambi,* which suggest the species does occur northward on the Pacific side of Middle America. The widely scattered records, however, indicate a now restricted distribution for the group.

Haffer (1967*a*, 1967*b*) has outlined the present distribution of nonforest vegetation in South America and has proposed dispersal routes based on the expansion and contraction of this biotype for the associated avifauna. His model also describes perfectly the zoogeographic relationships shown by all of the nonforest *Molossus* on this continent. *M. rufus, M. coibensis,* and *M. molossus,* for example, exist today isolated in the dry valleys of the upper Ríos Marañon, Huallaga, and Urabamba in Perú and must have been derived from a western Brazilian fauna that migrated northward along the Andean base during dry climatic periods when savannas and dry deciduous woodlands were advancing. If Haffer's thesis is correct, the Suriname and Guianan regions with their scattered xerophyllic vegetative enclaves is another filter zone for nonforest *Molossus*. Morphological data from this study for *M. molossus* substantiate this claim; a small bat occupies the northern Venezuelan coast west of the area but is replaced by a decidedly larger one southward in Brazil, which continues westward.

M. pretiosus has the most restricted and at the same time most disrupted distribution of all the mastiff bats. Populations in the dry upper reaches of the Magdalena and Cauca valleys of Colombia are well isolated from those in Central America. The species also occurs east of the Andean Cordillera Occidental in the broad llanos plain, but there are numerous records of populations beyond the southern terminus of this vegetative zone that surely represent colonies trapped in nonforest refugia by the retreating llanos. *M. pretiosus* has more or less supplanted *rufus* in Colombia, in all likelihood as a consequence of competitive exclusion.

The most puzzling zoogeographic pattern of all is shown by *M. molossus*. The forested filter zones in Panamá and the Guianan region probably account in large part for the broad array of morphotypes observed, but despite obvious morphological variation, genetic data link all together. Biochemical analyses further show that the Lesser Antillean populations were derived from the South American mainland. On the origin of the bats assigned to *M. molossus* in the Greater Antilles, I am less certain. Morphologically these bats are most like other populations referred to the species *molossus* yet specimens from Jamaica, as noted earlier, have a *sinaloae*-specific esterase allele. The genetic incongruity makes me suspect that a Middle American taxon, possibly *sinaloae*, gave rise at least to the Jamaican populations and perhaps those throughout the Greater Antilles. Pregill and Olson (1981) have described the overall distinctive nature of the Jamaican vertebrate fauna compared to the rest of the Greater Antilles and have proposed an invasion route across the submerged Gordo, Rosalinda, and Pedro banks east of Nicaragua and Honduras.

Phenotypic Expression

The genus *Molossus* is a compact group built on a highly uniform morphological theme that is exceeded only by *Cheiromeles* (Molossidae) for its degree of interspecific phenetic similarity (Freeman, 1981). Species are differentiated primarily on the basis of size, a fact attested to by the canonical variates analysis wherein almost 90 percent of all variation is defined by Vector I, the size-related component (see Figs. 2 and 3). *M. sinaloae* and *M. molossus* separated along Vector II due to narrower crania, which may parallel a diversification in food habits. Inasmuch as size is the best discriminator, it would appear that mastiff bats follow Schoener's (1965, 1968) ecological expectations of species differentiation first along a food axis, partitioning resources by size, and second along a habitat axis.

Intraspecifically, clinal trends in morphology were absent, being replaced instead by a confusing association of localities generally poorly related to geographic proximity. *M. molossus* in particular has been described as localized in occurrence and displaying significant mensural differences between samples within the same general geographic area (Jones *et al.*, 1971) and on the same island (Genoways *et al.*, 1981). Data reported here show that all species of *Molossus* apparently exist in numerous, morphologically discrete populations and can be considered polytypic. Strong interlocality differentiation also occurs in both *Eumops bonariensis* (Eger, 1977) and New World *Tadarida* (Carter, 1962) and may well represent a common molossid feature.

Pronounced morphological variation among populations of mammals is usually indicative of low vagility or strong territorial tendencies that act to reduce deme size and to promote differentiation through drift, as in the case of moles (Yates, 1978) and pocket gophers (Honeycutt and Schmidly, 1979). Social structuring (Chesser, 1982), environmental patchiness, and ethological barriers like homing behavior (Hedgecock, 1978) can produce the same effect as well. The precise mechanism or mechanisms operating among mastiff bats to yield a mosaic of morphotypes only can be speculated on because little is known of the ecology and behavior of this group. Several facts, though, are clear. *Molossus* is not prohibited from dispersing by any morphological impediment. The entire family Molossidae most definitely is adapted to fast, efficient, and enduring flight. Vaughan (1978) outlined the most salient structural features associated with molossid flight and provided examples of the taxon's spectacular capabilities by citing evidence of lengthy 50-mile foraging trips by guano bats and altitude records of 2000 feet for *Eumops perotis*. Some *Tadarida* annually migrate hundreds, and perhaps thousands, of miles. The genus *Molossus* exhibits a strong proclivity for specific types of watering sites. This preference for open, unobstructed approaches over undisturbed water may be dictated by the group's unique wing design. Smith and Starrett (1979) determined that molossids have by far the highest aspect ratio (length to width) among chiropterans. Narrow wings imply low lift capabilities. Molossids, therefore, must maintain high air speeds to remain aloft and consequently could be restricted in the places from which they can drink. Scattered occurrences of surface water with such unusual characteristics may disrupt species distribution.

Molossus is most often found occupying buildings with attics, although expansion joints beneath bridges also commonly serve as roosts. Prior to the arrival of man in Middle America, hollow trees probably formed the natural roost sites. At some point, *Molossus* seemingly shifted from trees to manmade structures. Inasmuch as human establishments in Middle America generally are situated along streams and rivers, roosting and watering opportunities afforded by scattered towns and villages might further act to subdivide species. Mastiff bats also are gregarious. The best estimates for colony size are provided in the field notes of D. C. Gall and R. W. Adams who collected in Central America with D. C. Carter during 1963. The largest gatherings observed were on the order of a thousand individuals, although some roosts contained only a few hundred. *M. molossus* may not form aggregations quite as large as other congeners, but this is only my impression. Males and females occupy one roost, but occasionally the ratio appears skewed in favor of one sex; juveniles and subadults often are collected in the company of adults.

Colonies give the appearance of being sedentary. In 1976, I revisited numerous localities first collected by D. C. Carter between 1963 and 1967 and discovered the same species still inhabiting the same buildings or utilizing the same drinking place. Consequently, I was able to compare morphological samples separated in time by 9–13 years. Means for *Molossus* caught in the sixties are indicated by either a star or a triangle in Figures 2 (males) and 3 (females). The 1975–76 populations to which these are being related are as follows: *M. rufus* triangle = 7, star = 1; *M. aztecus* 35, *M. coibensis* 19; *M. sinaloae* 31 (the sample of *M. rufus* denoted by a star was actually 55 kilometers north of population 1 but 1 represents the nearest known colony). Assuming mastiff bats are sedentary, a difference in morphology is to be expected after so many years. With the exception of *M. sinaloae,* this is precisely what was obtained. Population averages generally were displaced often to the extent that the 1960 sample defined a nonoverlapping set (male *coibensis* from Panamá; male *rufus* from El Salvador; female *rufus* from México; female *aztecus* from Guatemala). It is interesting how well samples separated by a decade mirrored one another; that is, although means differed, collections from the sixties were still more like those taken in the seventies than they were to those taken from other geographic localities. Small perturbations in form may reflect the low levels of genetic variability presented for stochastic sampling (average individual heterozygosity, $\bar{H}$, was 0.034, range 0.002–0.095) or the dampening effect on drift associated with colonies containing several hundred to a thousand bats. *Tadarida brasiliensis cynocephala* also has been described as sedentary, and anecdotal accounts indicate a reluctance by the colony to move any distance when its residence has been destroyed (Carter, 1962). Although its colonies number upwards of 5000 individuals, it too exhibits significant morphological variation among populations.

Chromosomally, the family Molossidae is extremely conservative. Of the 13 genera recognized by Freeman (1981), seven, representing approximately 26 species, have been karyotyped. The diploid number for all taxa is 48 with the exception of members of the genus *Molossops* and two species of *Eumops* (for a summary of molossid karyotypes see Warner *et al.,* 1974; Peterson and Nagorsen, 1975; Dulić and Mrakovčić, 1980). The preponderance of a 2n = 48 karyotype among Old and New World taxa presumably separated since the Miocene (Koopman, 1970) led Warner *et al.* (1974) to conclude that this count was probably ancestral for the family. Species of *Molossus* thus retain the primitive condition and show no evidence of gross chromosomal alteration during the last 25 to 30 million years. This is contrary to predictions of the chromosomal evolution model championed by Wilson *et al.* (1975) and Bush *et al.* (1977). According to their hypothesis, an absence of chromosomal variability is synonymous with a lack of population subdivision, which is necessary to promote rapid speciation and incorporation of rearrangements by inbreeding and drift. Yet both morphological and genetic data (discussed below) clearly suggest that mastiff bats exist in populations that engage in little genic exchange, that indeed are subdivided. The single karyotypic arrangement within *Molossus* is best explained as the end result of a selective process for the "optimum" karyotype for a molossid niche and as such is compatible with the canalization model of Bickham and Baker (1979).

Genetic Variation

If just those species of *Molossus* represented by more than three populations are considered (Table 9), an average level of differentiation among populations was 22 percent (F_{ST} = 0.12–0.47), with variation in most taxa ranging between 12 percent and 15 percent. Similar amounts of differentiation have been noted where gene flow is reduced as a consequence of low vagility, as among snails from different cities (16 percent, Selander and Kaufman, 1975) and house mice from different farms (17.3 percent, Selander, 1970). Chesser (1983) observed slightly higher values among coteries, the breeding units of prairie dogs (22 percent). Extremes of interpopulational variation among vertebrates have been recorded by Echelle *et al.* (1976) for darters (F_{ST} = 0.4–0.5), Loudenslager and Gall (1980) for cutthroat trout (F_{ST} = 0.45), Larson and Highton (1979) for plethodontid salamanders (F_{ST} = 0.70–0.77), and Patton and Yang (1977) for pocket gophers (F_{ST} = 0.41 for polymorphic loci only). Mechanisms restricting gene flow in the aforementioned studies are varied and range from competition among genera and geographic isolation to territoriality, small effective population size, and relatively low vagility.

Populations within species of *Molossus* exhibit low levels of genetic variability. The average individual was heterozygous at only three percent of the loci examined ($\bar{H}$ of 0.002 to 0.095). Half of the species were polymorphic at only five percent of the loci assayed (for example, *M. coibensis*, *M. molossus*, and *M. aztecus*); polymorphism among the remaining taxa often was due to extremely low frequency alleles (see Table 7). Maintaining genetic variation is necessary for species in order to avoid being channeled into an evolutionary "dead end." With a canalized karyotype, *Molossus* must rely heavily on crossing over and mutation to generate unique combinations. If the electrophoretic data are any indication of the relative homogeneity of the genome in mastiff bats, it would appear that the formation of chiasmata during meiosis is less critical than mutation for providing variability. Adopting a population structure that enhances contributions made by mutation would be advantageous to a group with monomorphic tendencies like *Molossus*.

Morphological and genetic data presented above indicate species in the genus *Molossus* exist in localized populations. A semi-isolated deme structure such as this sets the stage for differentiation by sampling drift. Novel genic combinations also increase more rapidly in small, inbreeding demes (Slatkin, 1976). The amount of drift, and hence its efficiency in furthering the development of interpopulational heterogeneity, is dependent on effective population size (N_e) and a balance between local inbreeding and immigration. According to Wright (1978), if inbreeding coefficients (F_{IS}) average 0.04 or greater, differentiation among populations is more likely to occur due to the imbalance of genetic drift and migration. This corresponds roughly to effective population sizes of less than 200 with the magnitude of drift inversely proportional to N_e. F_{IS} values determined for *Molossus* (Table 9) are sufficiently high to allow sampling error to produce genetic variability among populations. Inconsistent with this assumption is the observation that colonies of *Molossus* often number several hundred to a thousand individuals (see discussion above). Population size thus would appear too large to allow stochastic events to have an effect. I can only conclude in view of the morphological and genetic data (F_{IS} and F_{ST} values), which describe significant differentiation among populations, that the effective population actually is less than the number of bats occupying a roost. The means whereby the number of breeding adults is decreased is not known at present. Carter (1962) judged colonies of *Tadarida brasiliensis cynocephala* to contain between 5000 and 10,000 bats, yet he was able to identify significant mensural differences among populations. Differentiation was attributed to the sedentary nature of colonies, which suggests the effectiveness of isolation in promoting drift.

To maintain the degree of genetic difference observed among populations of *Molossus* (a constant F_{ST}, given that there are no selective forces acting), the number of dispersing individuals per generation must be less than two ($N_e M = (1/4F_{ST}) - 0.25$, where M is the dispersal rate; Ryman *et al.*, 1980). The estimated number of immigrants allowed per generation is shown in Table 9 for each species. Values for *M. pretiosus* and *M. bondae* were disregarded because only two populations were represented—a possible source of bias in assessing F_{ST}. *M. coibensis* from San Francisco, Panamá, offers evidence that exchange of individuals is uncommon and restricted (compare gene frequencies of EST-2 116 allele for Panamanian populations).

One obvious and generally negative consequence of inbreeding is fitness depression, a reduction in vigor presumably due to the random fixation of deleterious genes (Wright, 1980). Among some colonial mammals where social organizations have developed that would promote consanguineous matings, other mechanisms have evolved to counteract the effects of substructuring as among marmots (Schwartz and Armitage, 1980) and spear-nosed bats (McCracken and Bradbury, 1977). However, inbreeding can still be an effective strategy if the cost of dispersal is high, as is likely for prairie dogs (Chesser, 1983), or depression in fitness is low (Bengtsson, 1978). Inasmuch as the magnitude of inbreeding is a function of effective population size (assuming panmixia), *Molossus* may avoid the more severe implications of inbreeding by maintaining numbers in the hundreds and yet keeping densities low enough to take advantage of stochastic processes. Genetic variability is enhanced when populations are subdivided into partially isolated units susceptible to random differentiation (Wright, 1980). This may be a critical mechanism for enlarging the field of genetic variability in a relatively homogenic group like *Molossus*.

Molossus rufus É. Geoffroy St.-Hilaire

1805. *Molossus rufus* É. Geoffroy St.-Hilaire, Bull. Sci. Soc. Philom., 3(96):279 (error for 379)
1805. *Molossus castaneus* É. Geoffroy St.-Hilaire, Ann. Mus. Nat. Hist. Nat. Paris, 6:155.
1827. *Dysopes alecto* Temminck, Monographies de Mammalogie, 1:231, pls. 20, 23, figs. 23–26.
1843. *Dysopes albus* Wagner, Arch. Naturgesch., 9(1):368.
1844. *Molossus myosurus* Tschudi, Untersuch. Fauna Peruana, p. 83.
1891. *Molossus fluminensis* Lataste, Ann. Mus. Civ. Stor. Nat. Giacomo Doria, ser. 2, 10:658, 11 April.
1902. *Molossus nigricans* Miller, Proc. Acad. Nat. Sci. Philadelphia, 54:395, 12 September.
1955. *Cynomops malagai* Villa-R., Acta Zool. Mexicana, 1(4):2, 15 September.
1956. *Molossus pretiosus macdougalli* Goodwin, Amer. Mus. Novit., 1757:3, 8 March.

Type.—Adult male, in fluid with skull removed, MNHN A.428/224; Amérique (Cayenne, French Guiana, by restriction—Miller, 1913); collector and date of capture unknown. Lectotype.

Type material.—Two syntypes (MNHN A.428/224, A.428/224a) were located in the Muséum National d'Histoire Naturelle, Paris. Carter and Dolan (1978) designated A.428/224 as lectotype for the name *Molossus rufus* É. Geoffroy St.-Hilaire.

Distribution.—East of the Sierra Madre Oriental and west of the Sierra Madre Occidental in México. The two seemingly anomalous reports of specimens from Mexico City and Morelos (see Fig. 7) probably represent Pacific coastal invaders that entered the Balsas Basin by following the Río Mezcala. From México, known southward along the Caribbean versant as far as Honduras, and along the Pacific coast southward to Panamá and thence into South America (Figs. 8–9), where it has been taken only east of the Andes. Koopman (1978) considered Tschudi's (1844) record from Ceja, Arequipa, Perú, on the western Andean slope, as questionable.

M. rufus is associated with xeric thorn forests, savannas, and dry tropical deciduous forests (Leopold, 1952; Fleming, 1971) at elevations almost exclusively below 1000 meters, but the species has been collected occasionally at higher altitudes in parts of México, Costa Rica, and Perú. River systems appear to serve as important dispersal routes, permitting the species to occupy a larger geographic area than might otherwise be possible. However, mountain chains as barriers to movement and patchiness in areas of suitable habitat have coupled with a linear distribution along river courses and the proclivity of the species for a sedentary existence to subdivide populations into localized demes that are free to vary morphologically and genically from other neighboring populations.

Comparisons.—From *M. sinaloae*, *M. aztecus*, and *M. molossus*, *M. rufus* differs in being larger in most dimensions with a broader skull and stouter, thick-chested physique; having a short (approximately 2.0–2.5 mm.), black, uni-colored pelage rather than a long, bicolored one; possessing an unusually well-developed sagittal crest (Fig. 10), particularly in males; exhibiting a short, square muzzle as opposed to a more tapered nose; and bearing incisors that

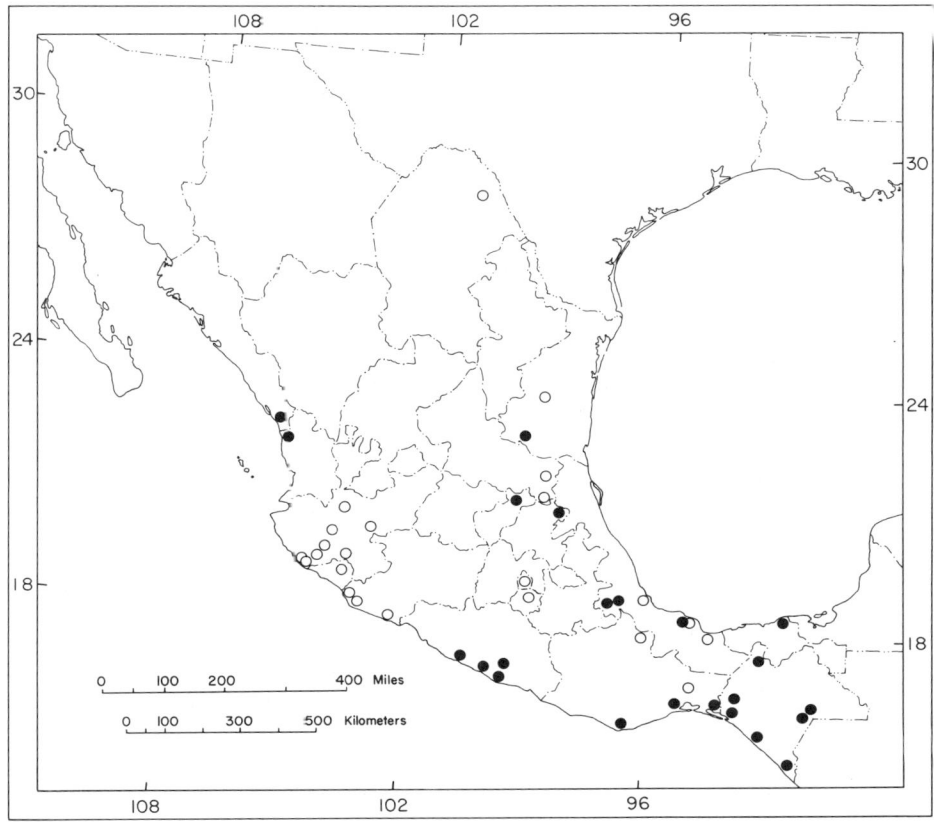

Fig. 7.—Distribution of *Molossus rufus* in México. Solid circles indicate specimens examined; open circles refer to literature records.

are spatulate rather than pincerlike (see Fig. 11). *M. rufus* resembles *M. pretiosus*, *M. bondae*, and *M. coibensis* in color and length of fur (although *coibensis* has more brown and less black in the hairs and the very base of individual hairs is cream to white in color), shape of skull, development of the sagittal crest, formation of the muzzle, and disposition of the incisors. It differs essentially only in size. Of all Central American *Molossus*, *M. rufus* is largest, but populations of this taxon on the Pacific coast between Guerrero, México, and La Libertad, El Salvador, average small for the species and approach, but seldom overlap, *M. pretiosus* in size. See Tables 3 and 4 for selected external and cranial measurements.

Remarks.—Goodwin (1960) proposed that the name *Molossus rufus* É. Geoffroy St.-Hilaire was a junior synonym of *Eumops auripendulus* (Shaw). His decision was predicated on the assumption that measurements presented by Geoffroy (1805a) in the original description of *rufus* were too large to represent the large species of *Molossus* then known from Central and South America and that the correct name for the large mainland form of *Molossus* was actually *M. ater.*

Goodwin seemingly ignored the content of Geoffroy's written and more detailed description of the taxon that appeared in his "Memoires" published the same year (Geoffroy, 1805*b*: 155). In it, *M. rufus* is identified as a true *Molossus* by the accurate description of the muzzle as "quite large and short." This contrasts sharply with Geoffroy's account of *M. ater* (which follows that of *rufus* on the same page): "Its muzzle is more tapered than is the preceding species [*Molossus rufus*]; its ears are perceptibly larger and particularly they are set farther back." The latter description suggests that *M. ater* Geoffroy is actually a *Eumops*. D. C. Carter has examined the two specimens located in the Paris Museum and labelled "*Molossus rufus* Geoff., Type A.428, Amerique" and has verified that they are indeed representatives of the genus *Molossus*. Despite Goodwin's misgivings concerning their authenticity, there is no reason to assume these specimens are other than what they are purported to be, types of *M. rufus* É. Geoffroy St.-Hilaire.

Measurements for the lectotype of *rufus* taken from Carter and Dolan (1978) and included in the MANOVA plot in Figure 2 clearly show it is morphologically indistinguishable from other large, Central American mastiffs. Given this strong physical similarity and the absence of any distinct patterns of clinal or geographic variation among Central American populations as well as the relative genic homogeneity shown over a wide geographic area, it seems prudent to recognize, at this time, a single taxon of large mastiff bat in the Western Hemisphere, *Molossus rufus*. Keeping in mind that distributional records indicate a hiatus between Middle and South American populations in the vicinity of Panamá and northern Colombia, future studies may show that subspecific status is warranted, in which case the name *nigricans* would apply to populations of *rufus* from Central America. Specimens collected from the type locality of *M. pretiosus macdougalli* during this study (population 14) failed to differ phenotypically from surrounding populations of *M. rufus* (see Figs. 2 and 3). Consequently, this name is considered a junior synonym of *rufus*.

Specimens examined (770).—Populations 1–14, 76, 82 (see Table 1). BRAZIL. Mato Grosso: Serra do Roncador, 284 km. N Xavantina, 10 (USNM). Minas Gerais: 3 mi. NE Viçosa, 8 (USNM). COSTA RICA. Alajuela: Concepción de Atenas, 21 (LACM). Puntarenas: 4 mi. NE Palmar (=Palmar Sur), 1 (TCWC); 3 mi. N, 4 mi. E Puntarenas, 1 (TCWC); 9 mi. ENE Puerto Golfito, 7 (TCWC); Rincón, 2 (LACM); Río Diciplina, 11.9 mi. N Palmar Norte, 3 (LACM); Villa Neilly, 3 (LACM). San José: Escazú, 1 (LACM); 4 mi. N San Isidro del General, 3 (TCWC). EL SALVADOR. Chalatenango: 20 km. W Chalatenango, 15 (TCWC). La Libertad: 0.8 mi. N, 9 mi. W La Libertad, 1 (TCWC); 3.5 km. N, 6.5 km. W Nueva San Salvador, 1 (TCWC). GUATEMALA. Alta Verapaz: Chinajá, 5 (KU). Izabal: 25 km. SSW Puerto Barrios, 6 (TCWC). Jutiapa: 6 mi. S Asunción Mita, 1 (KU). Petén: Toocog, 15 km. SE La Libertad, 2 (KU). Quezaltenango: 7 km. S, 9.5 km. E Coatepeque, 1 (TCWC). Retalhuleu: 12 km. SW Retalhuleu, 7 (TCWC); Retalhuleu, 11 (TCWC) Santa Rosa: Astillero, 3 (KU); 3 km. S, 6 km. E Cuilapa, 8 (TCWC). HONDURAS. Atlántida: 9 mi. W La Ceiba, 4 (TCWC); 17 mi. W La Ceiba, 4 (TCWC); 1 mi. W Tela, 1 (TCWC). Copán: Copán, 22 (TCWC). Cortés: La Lima, 8 (TCWC); San Pedro Sula, 7 (TCWC). La Paz: La Cruz Grande, 19 (AMNH). Ocotepeque: Nueva Ocotepeque, 17 (TCWC). MÉXICO. Chiapas: 5 mi. N Arriaga, 16 (TCWC); 3 mi. N Arriaga, 4 (TCWC); 3 mi. E Cintalapa, 29 (TCWC); 15 mi. SW Las Cruces, 12

(KU); 18 mi. S La Trinitaria, 29 (TCWC); Pichucalco, 22 (TCWC). Guerrero: 15 mi. NW Acapulco, 3 (AMNH); 2 mi. NW Acapulco, 1 (KU); El Papayo, 3 (TCWC); 5 mi. ESE Tecpan, 6 (KU); Tierra Colorado, 1 (TCWC); Tres Palos, 13 (TCWC). Hidalgo: 2 km. S, 2 km. W Huejutla, 8 (TTU). Oaxaca: Puerto Angel, 9 (KU); 4 mi. E (San Pedro) Tapanatepec, 8 (TCWC); 1 mi. S Tequisistlán, 15 (TCWC), 1 (KU). Querétaro: Jalpan, 5 (TCWC); 11 mi. NW Jalpan, 1 (TCWC). Sinaloa: Escuinapa, 88 (MSU). Tabasco: Frontera, 9 (MSU). Tamaulipas: 7 mi. W Ocampo, 1 (KU). Veracruz: 4 km. WNW Fortín, 2 (TCWC); Ojo de Agua del Río Atoyac, 6 (TTU); San Andrés Tuxtla, 4 (TCWC). Yucatán: Dzitas, 19 (MUS); Mérida, 6 (TTU). NICARAGUA. Chinandega: 4.5 km. N Cosigüina, 2 (KU); El Paraíso, 1 km. N Cosigüina, 17 (KU); Hacienda San Isidro, 2 (KU); Potosí, 21 (KU); San Antonio, 2 (KU); Hacienda Bellavista, Volcán Casita, 4 (KU). Madriz: 1 mi. SE Yalagüina, 1 (TCWC). Nueva Segovia: Corozo, 15 km. NNE Jalapa, 1 (KU); 3.5 km. S, 2 km. W Jalapa, 3 (KU). PANAMÁ. Chiriquí: La Concepción, 7 (TCWC); 11 mi. W La Concepción, 1 (TCWC); Progresso, 2 (USNM). Veraguas: 2 mi. S San Francisco, 1 (TCWC).

Additional records.—BRAZIL. São Paulo: Río de Janeiro (type of *M. fluminensis* Lataste, 1891). COLOMBIA. Arauca: Tame (Marinkelle and Cadena, 1972). COSTA RICA. Guanacaste: 4 km. NW Cañas, La Pacifica (LaVal and Fitch, 1977). Puntarenas: Boruca (Goodwin, 1946). MÉXICO. Coahuila: Morelos (Ramírez-Pulido and López-Forment, 1979). Colima: Colima (Miller, 1902). Distrito Federal: Mexico City (Villa-R., 1967). Jalisco (Watkins *et al.*, 1972, unless otherwise noted): Cihuatlán; Cuitzamala; El Grullo; El Zapote; 2 mi. N Tenacatita; Tenamastlán; Los Masos (J. A. Allen, 1906); Teuchitlán (Miller, 1902). Michoacán (Garrido-Rodríguez and López-Forment, personal communication, 1982, unless otherwise noted): Boca de Apiza (Villa-R., 1967); 48 km. S Coahuayana (Villa-R., 1967); Río Ostula, 11 km. SE La Placita; Punta San Telmo; Río Mexcalhuacán, 22 km. E Caleta de Campos by road. Morelos: 4 km. NW Coatlán del Río (Davis and Russell, 1954). Oaxaca (Villa-R., 1967): 6 mi. S Matías Romero; Tuxtepec. San Luis Potosí: Taninul (Villa-R., 1967); Río Moctezuma at Tamazunchale (Dalquest, 1954). Tamaulipas: 2 mi. S Ciudad Victoria (Davis, 1951). Veracruz: Catemaco (Miller, 1902); Paso de Ovejas (Villa-R., 1967); Miniatitlán (Villa-R., 1967). Yucatán: Chichén Itzá (Miller, 1902). PANAMÁ (Handley, 1966). Chiriquí; Alanje; 2 mi. W El Volcán. PARAGUAY. Villa Rica (=Villarrica) (Miller, 1913). PERÚ. Loreto: Boca Río Curaray (Koopman, 1978). Lagarto: Puerto Indiana; Sarayacu (Koopman, 1978). Amazonas: Pomara (Koopman, 1978). Cuzco: Marcapata (Tuttle, 1970). Huanuco: Tingo María; San Martín; Sapaja; Juan Guerra (Tuttle, 1970). SURINAME. Paramaribo (Husson, 1962). TRINIDAD (Goodwin and Greenhall, 1961).

Molossus pretiosus Miller

1902. *Molossus pretiosus* Miller, Proc. Acad. Nat. Sci. Philadelphia, 54:396, 12 September.

Type.—Adult male, skin and skull, USNM 102761; La Guaira, Distrito Federal, Venezuela; Wirt Robinson and M. W. Lyon Jr.; 13 July 1900. Holotype.

Type material.—Miller (1902) examined 71 specimens, in addition to the holotype, all from the vicinity of La Guiara. Twelve *Molossus* of medium size with forearm measurements ranging from 45.6–48.1 millimeters and collected on 6 July 1900 at La Guaira are extant at the United States National Museum (USNM 102737–48) and in all likelihood are surviving paratypes from Miller's original series.

Distribution.—Known along the Pacific versant from Nicaragua southward into South America (Figs. 8–9). In South America, the species has been recorded with certainty only from Colombia and Venezuela. The presence of *M. pretiosus* in the Central Andean valleys of Colombia is best explained as a series of relict populations that migrated up major river systems during drier climatic periods when nonforest biotypes were contiguous in this region.

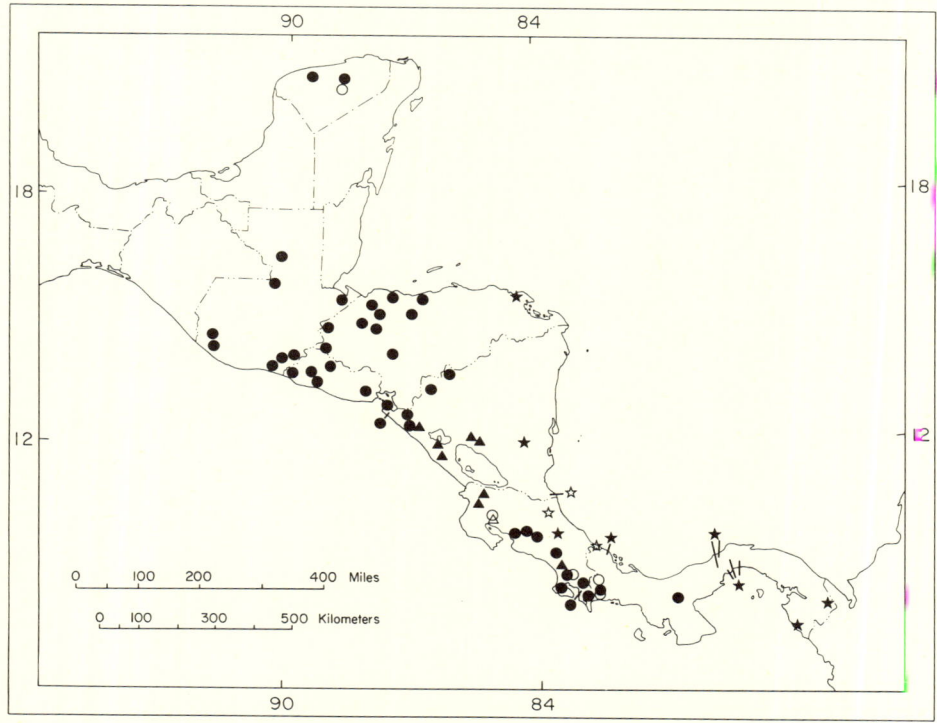

Fig. 8.—Distribution of *Molossus rufus* (circles), *M. pretiosus* (triangles), and *M. bondae* (stars) in Middle America. Solid symbols indicate specimens examined; open symbols refer to literature records.

Comparisons.—*M. pretiosus* strongly resembles *M. rufus* and *M. bondae* in all morphological characters (compare Figs. 10 and 11). It differs by being of intermediate size—larger than *bondae* but smaller than *rufus*. For characters differentiating *pretiosus* from all other *Molossus,* refer to the account for *M. rufus.* Older, worn pelages are a dark reddish brown to reddish orange color and are replaced by a deep, black coat during molt. See Tables 3 and 4 for selected external and cranial measurements.

Remarks.—*M. pretiosus* is remarkably similar to *M. rufus* (as defined here) and morphologically differs only in degree. Reluctance to accept *pretiosus* as a valid species over the years can be linked to this almost overwhelming mimicry and the scarcity of specimens to define a distribution and further test for the occurrence of two distinct morphotypes. An increased emphasis on field work during the last two decades has yielded the data to show conclusively (Figs. 8 and 9) that two size classes of large *Molossus* do exist (Jones *et al.,* 1971). Although the distribution of the smaller taxon, *pretiosus,* can be considered restricted when compared to some congeners (such as *rufus, molossus,* and *sinaloae*), it nonetheless covers a wide geographic range.

As noted in Table 7 and Figure 4, there were no fixed genic or chromosomal

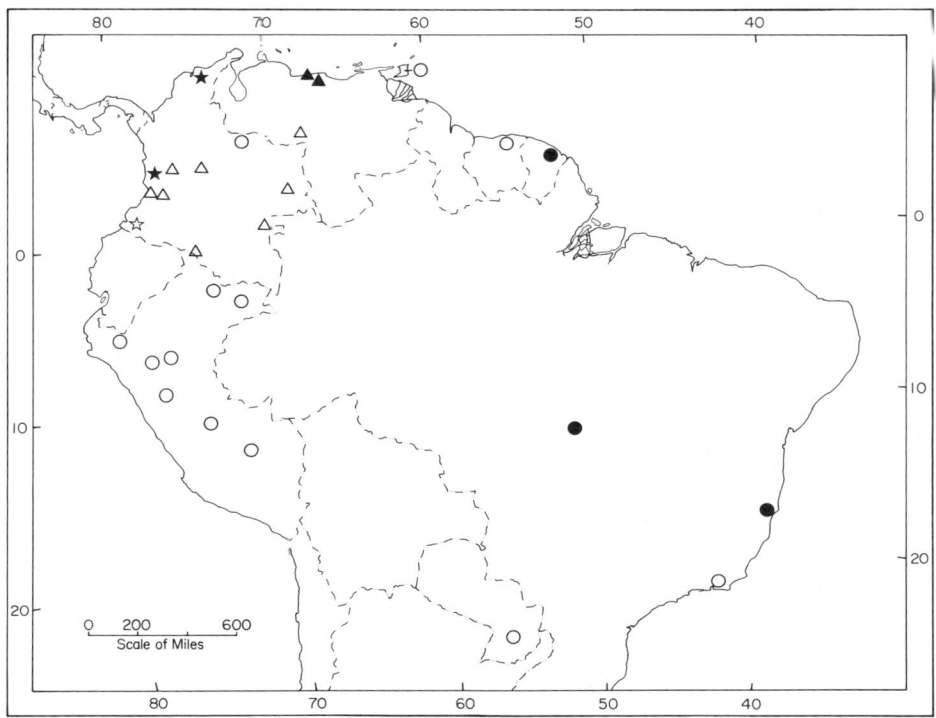

FIG. 9.—Distribution of *Molossus rufus* (circles), *M. pretiosus* (triangles), and *M. bondae* (stars) in South America. Solid symbols indicate specimens examined; open symbols refer to literature records.

differences detected between *pretiosus* and *rufus* that would argue in favor of considering them separate species. However, the obvious morphological integrity of the two taxa over a large geographic area in the absence of any obvious physical barriers to mixing and the location of the three points of apparent sympatry (San Antonio, Chinandega, Nicaragua; Río Diciplina, 11.9 mi. N Palmar Norte, Puntarenas, Costa Rica; La Pacifica, Guanacaste, Costa Rica) lead me to conclude that *M. pretiosus* is fulfilling the conditions of a species, reproductive isolation, which has been manifested by a distinct and consistent difference in size.

Although *M. pretiosus* has not been recorded from Panamá, it in all likelihood occurs there. Electrophoretic data available on a single specimen from Guatopo, Miranda, Venezuela (population 67), clearly unites South American *pretiosus* genetically with those in Central America (Table 7).

See the account for *M. rufus* for comments concerning the name *M. pretiosus macdougalli.*

Specimens examined (157).—Populations 15, 16, 68, 69 (See Table 1). COSTA RICA. Guanacaste: Cañas Dulces, 60 (TCWC). Puntarenas: 11.9 mi. N Palmar Norte, Río Diciplina, 1 (LACM). NICARAGUA. Boaco: 14 km. S Boaco, 28 (KU); 19 km. S, 2 km. E Boaco, 3 (KU). Carazo: 3 km. N, 4 km.

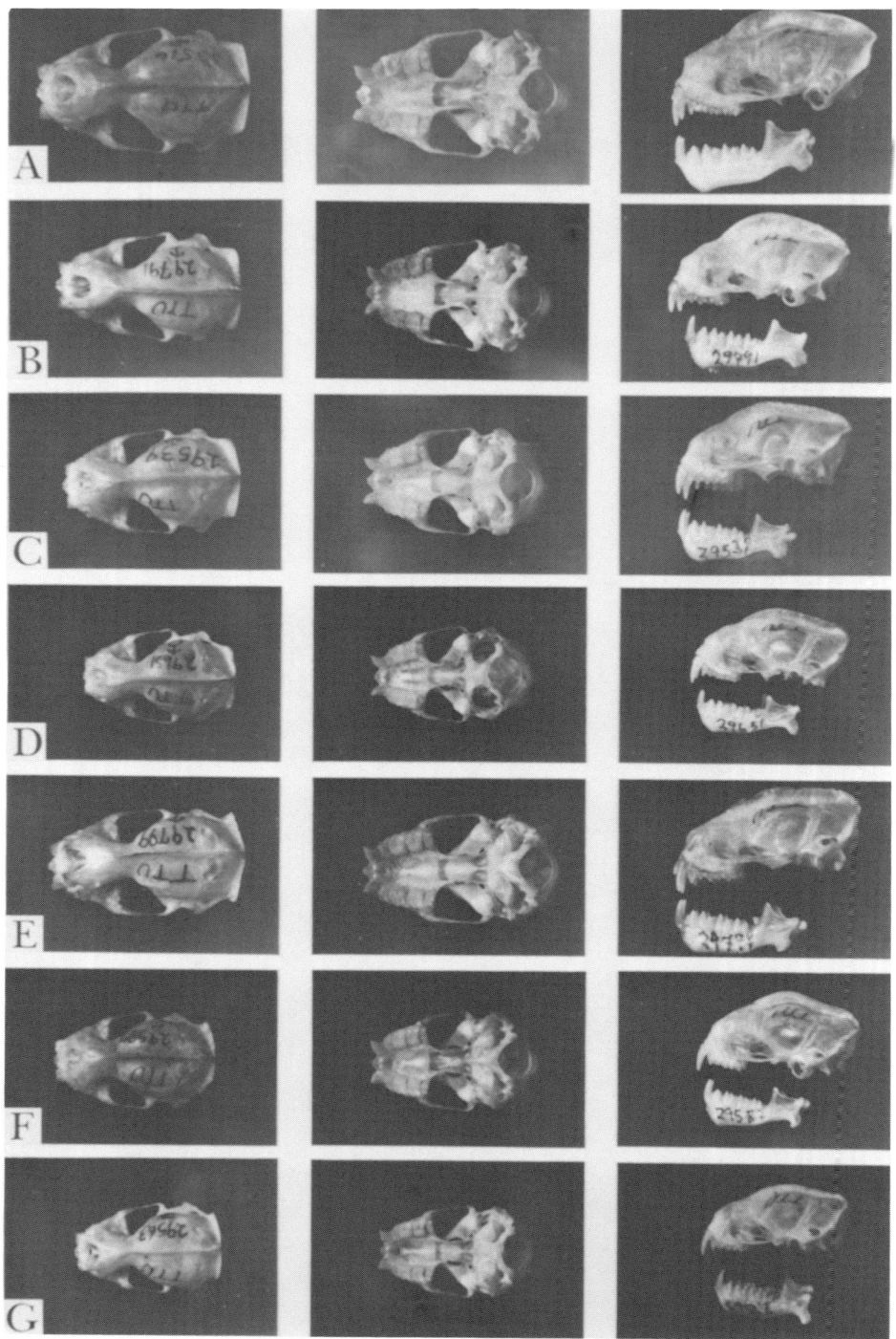

Fig. 10.—Dorsal, ventral, and lateral views of skulls for A, *M. rufus;* B, *M. pretiosus;* C, *M. bondae;* D, *M. coibensis;* E, *M. sinaloae;* F, *M. aztecus;* G, *M. molossus.*

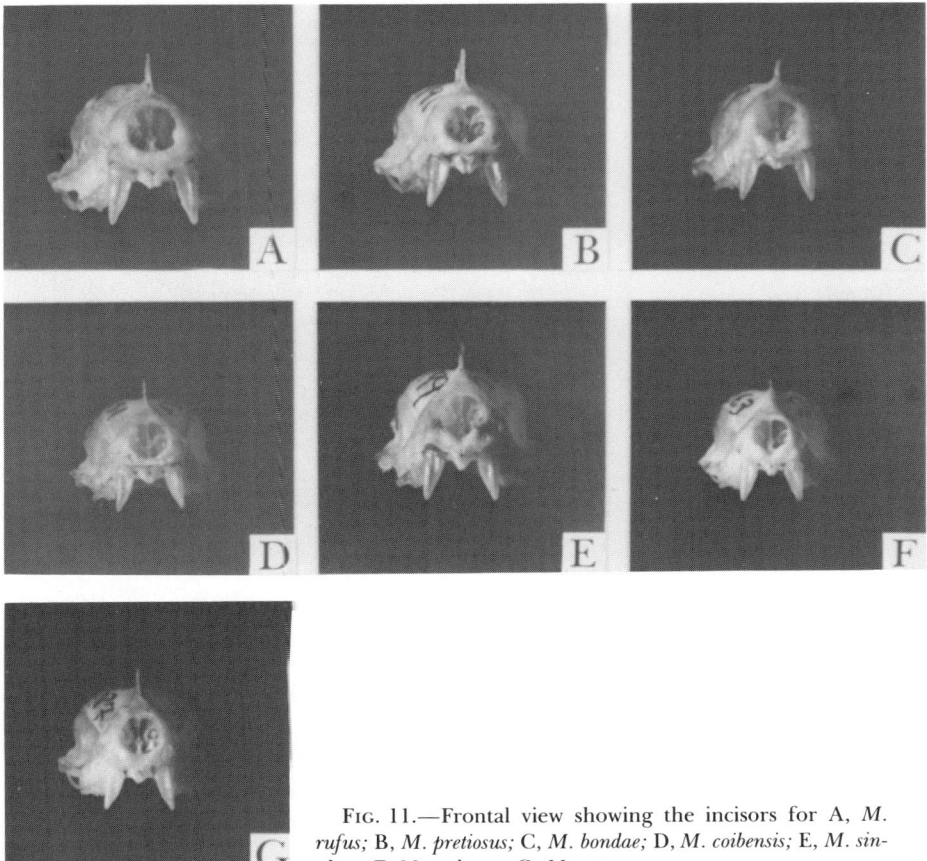

FIG. 11.—Frontal view showing the incisors for A, *M. rufus;* B, *M. pretiosus;* C, *M. bondae;* D, *M. coibensis;* E, *M. sinaloae;* F, *M. molossus;* G, *M. aztecus.*

W Diriamba, 21 (KU). Chinandega: San Antonio, 1 (KU). Managua: 6 mi. WSW Managua, 3 (KU).

Additional records.—COLOMBIA (Marinkelle and Cadena, 1972). Cundinamarca: San Juan de Río Seco. Guainía: Puerto Inírida. Putumayo: Puerto Leguizamo. Valle: Ansermanuevo; Cali; Cartago; Guabanal; Río Raposo. Vaupés: Mitú. COSTA RICA. Guanacaste: La Pacifica, 4 km. NW Las Cañas (LaVal, 1977). VENEZUELA. Apure: del nato El Frío (Ulargui, 1979).

Molossus bondae J. A. Allen

1904. *Molossus bondae* J. A. Allen, Bull. Amer. Mus. Nat. Hist., 20:228, 29 June.

Type.—Adult female, skin and skull, AMNH 23661; Bonda, Río Manzanares (7 mi. E Santa Marta), Magdalena, Colombia; Herbert H. Smith; 10 February 1900. Holotype.

Type material.—In addition to the holotype, Allen referred to three paratypes, in alcohol, collected by Francis C. Nicholas. Two of these are extant at the American Museum of Natural History.

Distribution.—In Central America (Fig. 8), restricted to the moist lowlands of the Caribbean versant at elevations below 3500 feet. Known with certainty from Brus Laguna, Honduras, southward to the Canal Zone in Panamá where it apparently "crossed over" to the Pacific Coast. Specimens from Honduras purported to be of this species by Goodwin (1942) were examined by LaVal (1977), who opined they were not referrable to *M. bondae*. A young female reported from Quintana Roo, México (Alvarez and Ramírez-Pulido, 1972), may actually prove to be *M. molossus*.

The distribution of this taxon in South America (Fig. 9) is obscure but it is seemingly restricted, possibly confined to Colombia inasmuch as reports of the species from other countries are lacking. Aellen (1970) summarized Colombian records but some specimens referred to in various papers are questionably *bondae*, like those reported from El Colegio in the Department of Cundinamarca (Valdivieso, 1964)—forearm measurements given are more representative of *coibensis* than *bondae*.

Comparisons.—*M. bondae* resembles *M. pretiosus* and *M. rufus* in every respect except size, it being the smallest of the three (compare Figs. 10 and 11). From all other species of *Molossus*, *M. bondae* differs in the same characters as *M. rufus* (see that account for details). Pelage color within a population has been observed to vary between a nonlustrous black and a reddish orange. Just as in the case of *M. rufus*, the reddish orange tint apparently results from wear. Selected external and cranial measurements appear in Tables 3 and 4.

Remarks.—*M. bondae* is a morphological mirror image of *M. rufus* and *M. pretiosus*, not known to occur sympatrically with either, and apparently confined to the east coast of Central America. Its close phylogenetic relationship to *rufus* and *pretiosus* is evident in its morphology and reflected in its genetic profile—a single detectable fixed difference at the LDH locus (see Table 7).

Specimens examined (175).—Populations 17, 18, 83 (see Table 1). COLOMBIA. Chaco: Condoto, 3 (BMNH). COSTA RICA: Cartago: Colorado, 32 (TCWC); Turrialba, 51 (LACM), 3 (TCWC). HONDURAS. Colón: Brus Laguna, 4 (TCWC). PANAMÁ: Bocas del Toro: Almirante, 4 (USNM); Sibube, 1 (USNM). Canal Zone: Chiva Chiva, 6 (USNM); Escobal, 3 (USNM); Ft. Kobbe, 8 (USNM); Ft. Sherman, 1 (USNM); Juan Mina, 1 (USNM). Darien: Jaqué, Río Imamado, 2 (USNM); Tacarcuna, 22 (USNM).

Additional records.—COLOMBIA. Chaco: Novita; Nariño: Barbacoas (J. A. Allen, 1916). COSTA RICA. Limón: Cariari, Río Tortuguero (Gardner *et al.*, 1970); Puerto Viejo (LaVal, 1977). NICARAGUA. Greytown (= San Juan del Norte) (Miller, 1913).

Molossus sinaloae J. A. Allen

1906.   *Molossus sinaloae* J. A. Allen, Bull. Amer. Mus. Nat. Hist., 22:236, 25 July.
1959.   *Molossus trinitatus* Goodwin, Amer. Mus. Novit., 1967:1, 29 October.

Type.—Adult female, skin and skull, AMNH 24524; Escuinapa, Sinaloa, México; J. H. Batty; 15 February 1904. Holotype.

Type material.—Allen examined no other specimens.

Distribution.—In México, stretching from the northern reaches of the Sierra Madre del Sur in Jalisco, southward and occurring eastward across the trans-

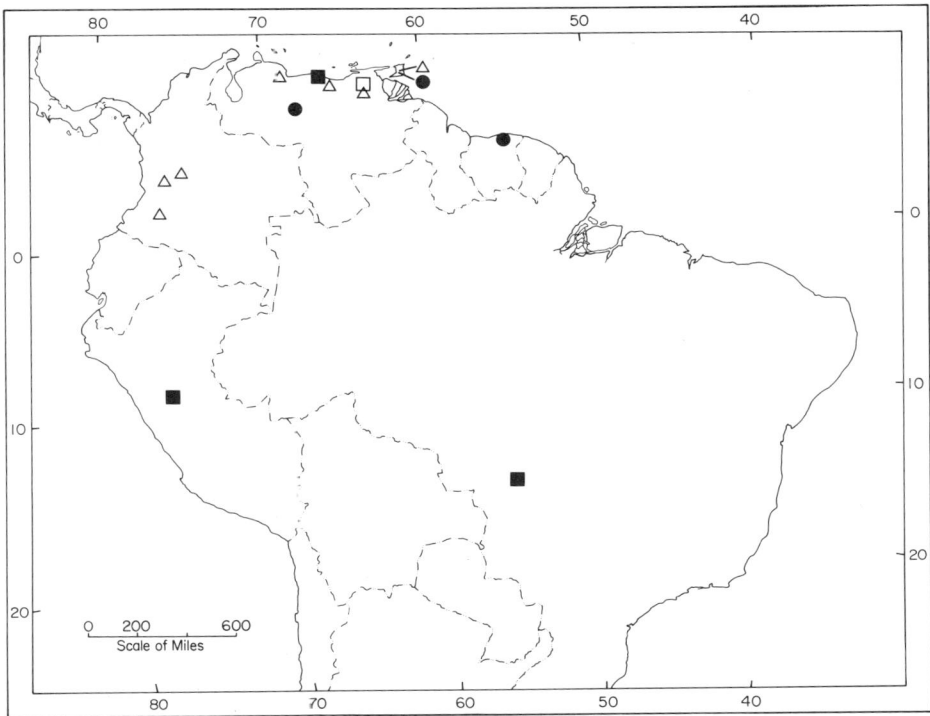

FIG. 12.—Distribution of *M. sinaloae* (triangles), *M. molossus* (circles), and *M. coibensis* (squares) in South America. Solid symbols indicate specimens examined; open symbols refer to literature records.

volcanic cordillera to Yucatán. Middle American specimens from Guatemala and Honduras present only north of the continental divide, but farther south the taxon occurs along both the Pacific and Caribbean versants. Literature records for South America limit *M. sinaloae* to the Central Andean valleys of Colombia and the coastal lowlands of northern Venezuela. See Figures 12 through 14. With the exception of its supposed occurrence at Escuinapa (the type locality), data available indicate that this species occurs exclusively in mesic tropical habitats. See the section on remarks for a discussion of the type locality for *M. sinaloae*.

Comparisons.—For characters differentiating *M. sinaloae* from *M. rufus, M. pretiosus,* and *M. bondae,* refer to the account for *rufus* (also compare Figs. 10 and 11). Traits separating *sinaloae* from *coibensis, aztecus,* and *molossus* are given under each of those species comparisons. The pelage is a dull, dark brown with a hint of red. See Tables 3 and 4 for selected external and cranial measurements.

Remarks.—Genetically, *M. sinaloae* is the most distinctive Middle American *Molossus* examined with species-specific alleles at the EST-2, α-GPD, and MDH loci. One peculiarity of its genetic make-up that remains to be explained is the

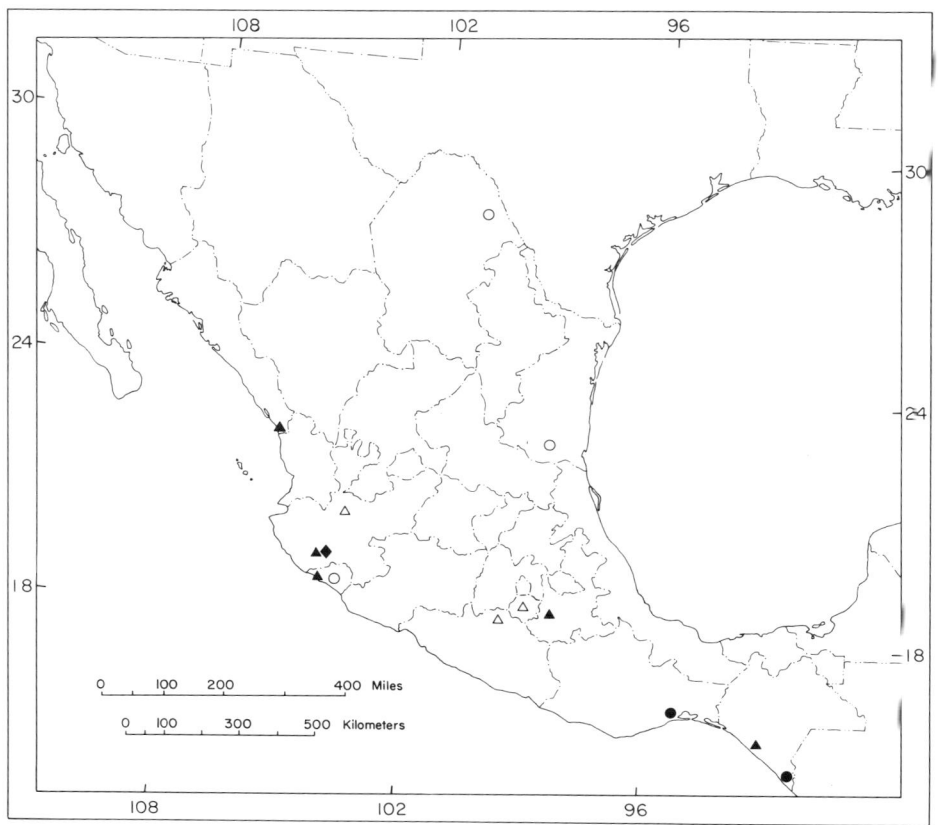

Fig. 13.—Distribution of *M. sinaloae* (triangles), *M. aztecus* (diamond), and *M. molossus* (circles) in México. Solid symbols indicate specimens examined; open symbols refer to literature records.

presence of an otherwise unique EST-2 108 allele in populations of what appear to be *M. molossus* on Jamaica.

Although generally separated by habitat differences, *M. sinaloae* and *M. rufus* have been taken sympatrically at three localities in Honduras where dry savannalike environments occur in proximity to more mesic vegatative zones. *M. aztecus* and *M. bondae*, which show habitat preferences similar to those exhibited by *sinaloae*, have been taken together with that taxon.

As noted earlier by Dolan and Carter (1979), an interesting paradox exists between the type locality reported for *M. sinaloae* and the known habitat preferences of that species. The vicinity surrounding Escuinapa, Sinaloa, is decidedly arid, consisting of sandy knolls dotted with thorny bushes, yucca and grass, altogether unlike the relatively mesic tropical forests from which all other specimens of *sinaloae* I have examined have been collected. Because all subsequent attempts to locate this bat in Sinaloa have failed, one is tempted to

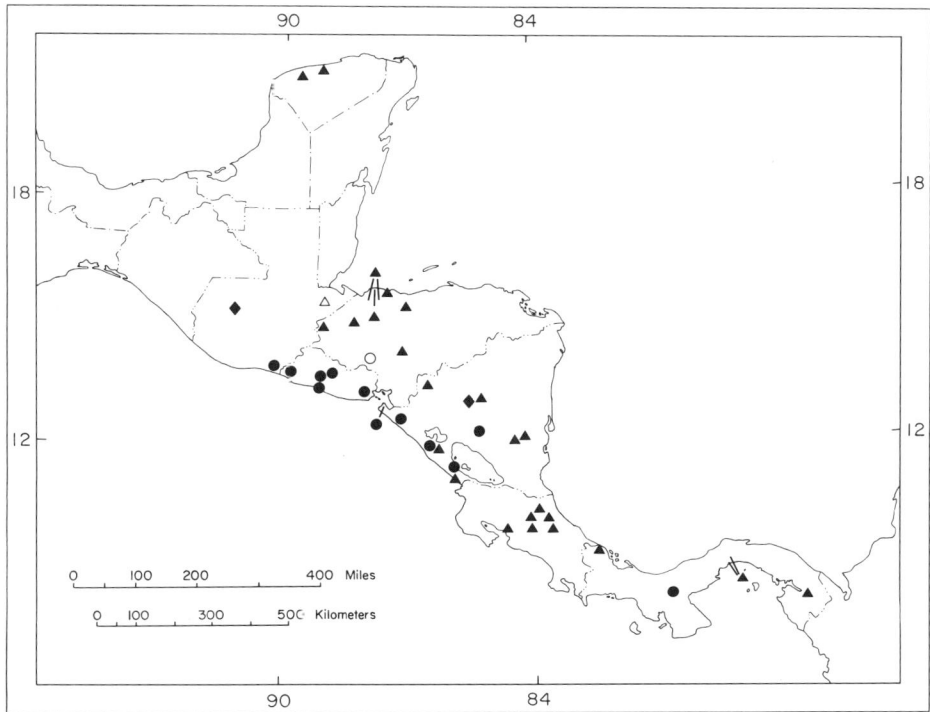

FIG. 14.—Distribution of *M. sinaloae* (triangles), *M. aztecus* (diamonds), and *M. molossus* (circles) in Middle America. Solid symbols indicate specimens examined; open symbols refer to literature records.

doubt the accuracy of the information reported for the type locality and to suspect a possible mix up in the field by J. H. Batty.

K. F. Koopman recently checked AMNH catalogs and the original field notes made by Batty and assures me this is extremely improbable. The holotype for *sinaloae* was taken on 15 February 1904 and received by the American Museum on 15 June of that same year. The Museum thus had the type in their possession five months before Batty broke camp in the vicinity of Escuinapa (10 November) and relocated in Jalisco (6 December). I can only conclude that the record for Escuinapa is real but imagine that the bat collected there represents a lost individual and not a member of any viable colony—*M. rufus* is the common member of the genus within that region. Due to the foregoing, I have elected to confine the description of the northern limit of the distribution of the species to that area in which we know it to occur today—the northern expanse of the Sierra Madre del Sur, specifically Teuchitlán, Jalisco (Watkins *et al.*, 1972).

Although Goodwin (1959) conferred specific status on specimens from Trinidad, and Freeman (1981) continued to acknowledge this ranking, a better understanding of the patterns of molossid variation make me disinclined to agree. The position of the holotype of *trinitatus* clearly within the *sinaloae* cluster (Fig. 2) argues against specific recognition. Whether *trinitatus* represents an identifiable subspecies, as suggested by Ojasti and Linares (1971), is still open to conjecture.

The skull of the holotype was unavailable to D. C. Carter, and consequently the type had to be excluded from the MANOVA.

Specimens examined (307).—Populations 24–31, 43, 86 (see Table 1). Costa Rica. Alajuela: Carblanco, 19 (TCWC). Cartago: Turrialba, 1 (TCWC). Heredia: Puerto Viejo, 5 (KU). Limón: Los Diamantes, 2 (LACM). Puntarenas: 6.5 mi. N, 2 mi. W Puntarenas, 1 (TCWC). San José: Escazú, 7 (LACM). Honduras. Atlántida: Tela, 2 (TCWC). Copán: Copán, 1 (TCWC). Cortés: La Lima, 1 (KU); 1 mi. W La Lima, 2 (TCWC), 4 (KU); NE end Lake Yojoa, 36 (TCWC). Francisco Morazán: 10 mi. N Talanga, 3 (TCWC). Santa Bárbara: Santa Bárbara, 12 (TCWC). Yoro: El Progreso, 2 (TCWC); Yoro, 39 (TCWC). México. Colima: 2 km. E Santiago, 1 (KU). Jalisco: 5 mi. S El Grullo, 5 (KU). Puebla: 2 mi. SE Izúcar de Matamoros, 5 (KU). Yucatán: 66 km. NE Mérida, 1 (KU). Nicaragua. Madriz: 1 mi. SE Yalagüina, 1 (TCWC). Managua: 3 mi. SW Managua, 1 (KU); 8 km. SW Managua, 2 (KU). Rivas: San Juan del Sur, 1 (KU). Zelaya: El Recreo, S side Río Mico, 28 (KU); 6 mi. W Rama, 43 (TCWC). Panamá. Boca del Toro: 7 km. SSE Chanquinola, 5 (USNM). Canal Zone: Ft. Amador, 1 (USNM); Ft. Clayton, 3 (USNM). Darien: Boca de Cupe, 1 (USNM); El Real, 1 (USNM).

Additional records.—Colombia (Marinkelle and Cadena, 1972). Cauca: Mina California (near Tambo). Tolima: Honda. Valle: Ansermanuevo. Guatemala. Izabal: Bobos (Jones, 1966). México. Guerrero: 16 km. E Teloloapan (Villa-R., 1967). Morelos: Jiutepec (Villa-R., 1967). Jalisco: Teuchitlán (Watkins *et al.*, 1972). Trinidad. Belmont: Port-of-Spain (Goodwin, 1959). Venezuela (Handley, 1976). Miranda: 1 km. S Río Chico. Monagas: San Agustín, 5 km. NW Caripe. Yaracuy: 10 km. NW Urama.

Molossus aztecus Saussure

1860.—*M[olossus]. aztecus* Saussure, Rev. Mag. Zoologie Paris, ser. 2, 12:285, pl. 15, fig. 3, 3a, July.

Type.—Adult of undetermined sex, skin and skull, MNHN 516.15; Amecameca ("at the foot of Popocatepetl"), Tlaxcala, México; collector and date of capture unknown. Holotype.

Type material.—Suassure referred to only a single specimen in his description.

Distribution.—Upland, generally mesic habitats at elevations above 1500 feet from the northernmost reaches of the Sierra Madre del Sur and transvolcanic cordillera in México southward to the central highlands of Nicaragua (Figs. 13 and 14). It is possible that *M. aztecus* is endemic to these cool upland regions and ranges no farther south than the Sierra de Amerique in Nicaragua, where the Central American uplift terminates and is discontinuous with the Costa Rican and Panamanian cordilleran chain.

Comparisons.—From *M. molossus, M. aztecus* differs in having a shorter forearm but larger skull and broader rostrum; hairs white along basal one-quarter or only at point of insertion as opposed to basal half of hair being white, as in *M. molossus;* incisors more spatulate than pincerlike (Fig. 11); and color generally deep chocolate brown as compared to a paler toffee brown in Central American specimens of *molossus*. A reddish-tinted pelage has not been recorded, which probably is a reflection of the paucity of specimens available. For characters differentiating *aztecus* from *sinaloae, coibensis,* and the *rufus-*complex, refer to accounts for those taxa. See Tables 3 and 4 for selected external and cranial measurements.

Remarks.—Although difficult to separate from *M. molossus* when alive, *M. aztecus* possesses a suite of mensural characters that easily identify it in a multivariate morphological analysis (Figs. 2 and 3). Genetically, *aztecus* also is readily identifiable, exhibiting fixed differences, when compared to *molossus,* at four loci (α-GPD, IPO-2, PGM-1, IPO-3) as well as an unique esterase allele (see Table 7). In terms of its genetic make-up, pelage type, and cranial configuration, *aztecus* appears to be more like *coibensis* than *molossus.*

Three clearly recognizable taxa of small *Molossus* with white-based hairs inhabit Central America; the smallest of these is *coibensis.* Of the two remaining taxa, one has been taken only in the lowlands, the other exclusively in upland areas. The specific epithet, *aztecus,* has been applied here to the montane species because of the elevation at which the holotype was captured, approximately 8000 feet. The forearm measurement of 36 millimeters reported by Saussure (1860) coincides with values obtained for other females from the Central American highlands and suggests that the holotype may have been of that sex.

Although the species is unknown from the highlands of Costa Rica and Panamá, I have examined a single female specimen from 9 mi. S Zaruma, El Oro, Ecuador, elevation 2000 feet, that might be assignable to *Molossus aztecus.* It agrees with those from the Middle American highlands in the length and color of pelage and cranial size but differs by having a greater forearm length of 39.6 millimeters. However, it is impossible at the present time to resolve the status of *M. aztecus* in South America.

Specimens examined (56).—Populations 32, 35, 39 (see Table 1). GUATEMALA: Huehuetenango. 1 km. NE Aguacatán, 28 (TCWC). MÉXICO. Jalisco: 5 mi. S El Grullo, 2 (KU). NICARAGUA. Matagalpa: 6 km. N El Tuma, 19 (TCWC).

Additional records.—Cranial and external measurements for a number of other specimens suggest they might be referrable to *aztecus* but positive assignment cannot be made until the configuration of the upper incisors is checked. The localities from whence this material comes are listed here although they have been excluded from the distribution maps. MÉXICO. Jalisco: 10 mi. NNE Pihuamo (KU); 16 mi. NE Tamazula (TCWC). Oaxaca: 20 mi. S, 5 mi. E Sola de Vega (KU).

Molossus coibensis J. A. Allen

1904. *Molossus coibensis* J. A. Allen, Bull. Amer. Mus. Nat. Hist., 20:227, 29 June.
1905. *Molossus barnesi* Thomas, Ann. Mag. Nat. Hist., ser. 7, 15:584, June.
1916. *Molossus cherriei* J. A. Allen, Bull. Amer. Mus. Nat. Hist., 35:529, 24 July.
1966. *Molossus aztecus lambi* Gardner, Los Angeles Co. Mus. Contrib. Sci., 111:1, 9 November.

Type.—Adult male, skin and skull, AMNH 18731; Coiba Island, Panamá; J. H. Batty; 3 June 1901. Holotype.

Type material.—Three additional specimens from the type locality were examined by Allen but were not referred to by number in the original description. These paratypes were part of a large Panamanian collection made by

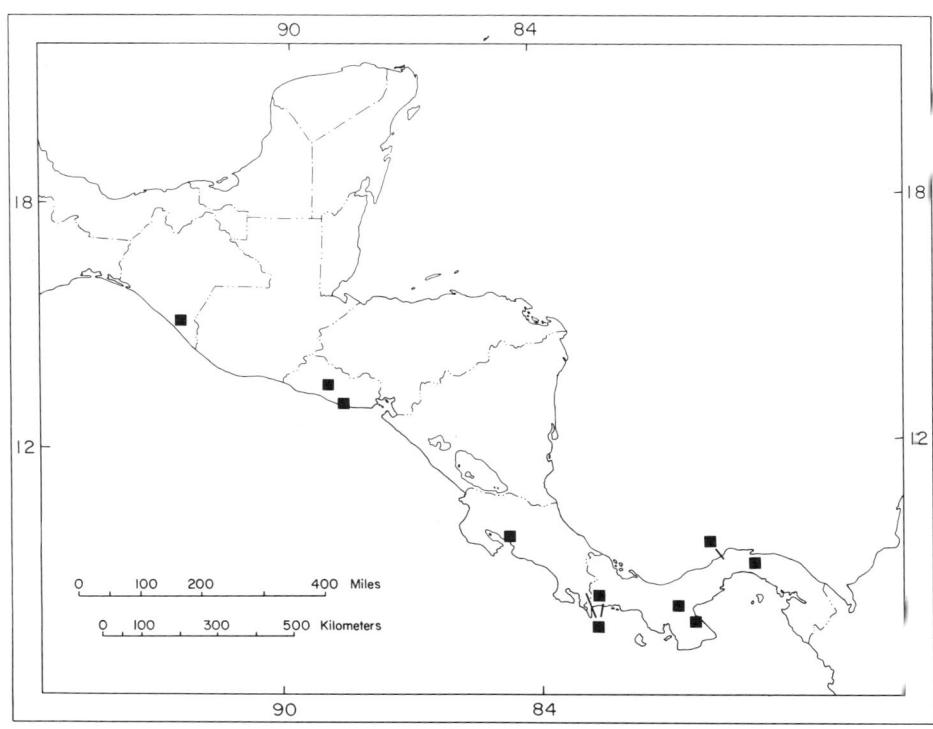

Fig. 15.—Distribution of *M. coibensis* in Central America. Symbols denote specimens examined.

Batty while in the employ of the American Museum and probably still exist at that institution.

Distribution.—Currently recorded only from the Pacific coast of Central America from Chiapas, México, southward (Fig. 15). In South America, the distribution is essentially unknown (Fig. 12). *M. coibensis* exhibits habitat preferences similar to *M. rufus,* that is dry environs at low elevations. All Middle American records of this taxon are from elevations below the 1000-foot contour line, but the Peruvian specimens examined in this study were caught at 2500 feet.

Comparisons.—From *M. rufus, M. pretiosus,* and *M. bondae, M. coibensis* differs in being smallest and in having a band that is cream to white in color at the base of the dorsal hairs. Compared to *M. sinaloae* and *M. molossus, M. coibensis* has a shorter forearm, shorter dorsal pelage (only 2–3 mm.); upper incisors that are spatulate in shape (Fig. 11), and a smaller band of white at the base of the dorsal hairs (band is one-quarter or less of the hair length in *coibensis,* one-quarter to one-half in *sinaloae* and *molossus*). Both the sagittal and lamboidal crests are better developed in *coibensis* (Fig. 10) than in either *aztecus* or *molossus;* the pectoral region of *coibensis* is also more robust than that of *molossus.* The fur is primarily deep black in color but grades into chocolate brown as the pelage wears. A unique allele at the LDH locus further distinguishes *coibensis* from all other species of *Molossus*. See Tables 3 and 4 for selected external and cranial measurements.

Remarks.—Morophologically, *M. coibensis* appears as an extreme miniaturization of the *rufus* morphotype and is clearly the smallest species present in Central America. Freeman's (1981) suggestion of conspecificity between *bondae* and *coibensis* notwithstanding, the specific status of *coibensis* is confirmed here by electrophoretic data that show the taxon carries a unique LDH allele and by two records of sympatry with the morphologically similar *M. molossus* at San Francisco, Panamá Some confusion does arise in attempting to differentiate *coibensis* from specimens of *M. molossus* in certain localities in South America, such as Venezuela, where the latter tends to be small. As a consequence, certain individual external measurements, like those of the forearm, converge with those of *coibensis*. However, cranial comparisons easily differentiate them: the skull of Panamanian *coibensis*, relative to small Venezuelan *molossus*, is larger with a broader, shorter rostrum and a more inflated, less elongated braincase; the upper incisors form the relatively flattened bladelike edge at their distal tip, as mentioned above, and do not taper into a set of pincers as in *molossus* (compare D and F in Fig. 11).

Sufficient cranial and external measurements were available for the holotypes of *M. barnesi*, *M. aztecus lambi*, and *M. coibensis* to include them in the MANOVA analysis (Figs. 2 and 3), and results clearly indicate that the aforementioned types bear a striking resemblance to one another. For this reason, these names are considered here to be synonyms. J. E. Hill of the British Museum graciously compared Panamanian *coibensis* collected during this investigation with the holotype of *M. barnesi* and noted no difference in the construction of the upper incisors (personal communication). However, based on variation in the extent of the white basal band in the dorsal fur, features of the basisphenoid pits, breadth of the mesopterygoid canal, and the absence of geographically intermediate populations, Hill argued for continued recognition of *M. barnesi*. In the case of *M. cherriei*, the skull has been lost, but all other details presented by Allen in the original description regarding size of the forearm, pelage length and coloration, and presence of a minute white band at the base of the hairs identify *cherriei* as a junior synonym of *coibensis*.

Specimens examined (172).—Populations 19–23, 50, 56, 58, 71, 77 (see Table 1). BRAZIL. Matto Grosso: Tapirapoan, 1 (AMNH) type for *M. cherriei*. COSTA RICA. Puntarenas: Boca de Barranca, 5 (LACM); Puerto Cortés, 4 (LACM); Villa Neily, 11 (LACM). EL SALVADOR. La Paz: 3 mi. NW La Herradura, 2 (TTU). San Salvador: San Salvador, 1 (SMF). PANAMÁ. Chiriquí: La Concepción, 36 (TCWC).

Additional records.—VENEZUELA. Monagas: Caicara (BMNH personal communication from J. E. Hill).

Molossus molossus (Pallas)

1766.   *V[espertilio]. Molossus* Pallas, Miscellanea Zoologica, 49.
1792.   *V[espertilio]. mol[ossus]. minor* Kerr, The animal kingdom . . . , p. 97.
1792.   *V[espertilio]. mol[ossus]. major* Kerr, The animal kingdom . . . , p. 97.

1805. *Molossus obscurus* É. Geoffroy St.-Hilaire, Ann. Mus. Nat. Hist. Nat. Paris, 6:155.
1805. *Molossus longicaudatus* É. Geoffroy St.-Hilaire, Ann. Mus. Nat. Hist. Nat. Paris, 6:155.
1805. *Molossus fusciventer* É. Geoffroy St.-Hilaire, Ann. Mus. Nat. Hist. Nat. Paris, 6:155.
1838. *Molossus fuliginosus* Gray, Mag. Zool. Bot., 2:501.
1839. *Molossus tropidorhynchus* Gray, Ann. Nat. Hist., ser. 1, 4:6, September.
1900. *Molossus pygmaeus* Miller, Proc. Biol. Soc. Washington, 13:162, 31 October.
1908. *Molossus verrilli* J. A. Allen, Bull. Amer. Mus. Nat. Hist., 24:581, 11 September.
1913. *Molossus debilis* Miller, Proc. U.S. Nat. Mus., 46:90, 23 August.
1913. *Molossus fortis* Miller, Proc. U.S. Nat. Mus., 46:89, 23 August.
1916. *Molossus daulensis* J. A. Allen, Bull. Amer. Mus. Nat. Hist., 35:530.
1952. *Molossus milleri* Johnson, Proc. Biol. Soc. Washington, 65:197, 5 November (replacement name for *M. fuliginosus* Gray, 1838).

Type.—Adult male, in fluid, MNHN A.419/225 (also plate XIX, fig. 1 of Buffon and Daubenton, 1763); Martinique, Lesser Antilles, by restriction (Husson, 162); collector and date of capture unknown. Lectotype.

Type material.—The name *Vespertilio molossus* was based on a composite series containing one specimen of *Tadarida* and the two bats figured in plate XIX of Buffon and Daubenton (1763). The larger bat in figure 1 has been designated lectotype by Husson (1962); the smaller bat in figure 2 is an extant specimen at the Paris Museum, MNHN 792, which is also the holotype for *M. longicaudatus* É. Geoffroy St.-Hilaire.

Distribution.—Known with certainty to occur on the Pacific coast from Oaxaca, México, to Rivas, Nicaragua (Figs. 13 and 14). The taxon is as yet unreported from Costa Rica and the Caribbean versant of Middle America (however, see the section on remarks for a record by Murie, 1935). *M. molossus* also has been taken in Panamá (Fig. 14) and is known throughout the Greater and Lesser Antillean Islands. On the mainland of South America (Fig. 12), the taxon occupies a narrow zone along the northern coast and has been reported primarily from the western slope of the Andes in Perú (see Koopman, 1978). A larger bat apparently dominates the continental interior from the occidental Andean hills in Colombia, Ecuador, and Perú eastward to Surinam and at least as far south as Argentina.

In Middle America, *M. aztecus* and *M. molossus* appear to be separated altitudinally. It is on this premise that specimens reported from the lowland areas by Villa-R. (1967) in Colima, Chiapas, and Tamaulipas, and by Ramírez-Pulido and López-Forment (1979) for Coahuila tentatively are assigned here to *M. molossus*. Specimens reported by Watkins *et al.* (1972) from Jalisco are all from upland situations and in all likelihood represent *M. aztecus*.

Comparisons.—For characters differentiating *M. molossus* from *M. rufus*, *M. pretiosus*, and *M. bondae*, refer to the account for *rufus* (also compare Figs. 10 and 11). Characters separating *molossus* from *coibensis*, *sinaloae*, and *aztecus* are given under each of those species comparisons. Color is somewhat variable with geographic location. In Central America, *M. molossus* is a toffee brown. See Tables 3 and 4 for selected external and cranial measurements.

Remarks.—The tortuous history of the species name *molossus* has been outlined in detail by Husson (1962), but a few additional comments need to be made. Briefly, the binomen *Vespertilio molossus* first was applied by Pallas (1766) to a composite group of specimens: a large bat (fig. 1) and small bat (fig. 2) illustrated by DeSeve in Daubenton's article in Buffon and Daubenton (1763) and at least one actual specimen of *Tadarida*. To avoid the complicated, and unacceptable, condition of the type species of the genus *Molossus* being a member of the genus *Tadarida*, which would be the consequence of selecting as type Pallas' specimen from "America," Husson designated as lectotype the larger of the two bats figured by Daubenton. The type locality has been taken as "Martinique" based on a statement to that effect made by Daubenton in another publication in which he first described this specimen. Although Daubenton (Buffon and Daubenton, 1763:87) referred to this description in a footnote as appearing in the "Memoires de l'Academie royale des Sciences, annee 1759," Husson found the paper in a periodical entitled "Histoire de l'Academie royale des Sciences. Annee 1759. Avec les Memoires de physique, pour la meme Annee. Tires des registres de cette Academie. Tome troisieme" and noted that the title of the article was "Memoire sur les chauve-souris" with the bat in question being discussed on pages 111–112. This paper apparently was not actually printed until 1777. Miller (1913) located the description in what he referred to as "Memoires of the Royal Academy of Sciences, Paris, for the year 1759" but gave 387 as the page on which the "mulot volant" of figure 1 was referenced and 1765 as the year of publication. It seems likely that the publication cited by Miller is merely a literal translation of that given in Daubenton's footnote (Buffon and Daubenton, 1763:87) and not an accurate title because no such periodical occurs in the rather extensive listing of serial publications published by the British Museum (Natural History), 1975. Whatever the periodical, it is clear from the different pagination and publication date that the reference seen by Miller was not the same as that of Husson (1962) and that the description of Daubenton's "mulot volant" appeared in no less than three different articles. Although Daubenton failed to identify the place of origin of the "mulot volant" in his 1763 work, he consistently referred to it as having come from Martinique in the other two papers.

Three "small" *Molossus* exist today at the Paris Museum; two of almost the same size are preserved in fluid and one smaller bat is preserved as a skin and skull. This is the same number held by Daubenton, and I suspect it is his original series. The small bat represented by the skin and skull, MNHN 792, is the holotype of *M. longicaudatus* É. Geoffroy, and according to Geoffroy is the specimen portrayed in figure 2, plate XIX (bat no. DCDVII), of Buffon and Daubenton (1763). This appears accurate because forearm measurements taken from the figure (36 mm.) are almost precisely those recorded for *longicaudatus* (36.9 mm.) by Carter and Dolan (1978). Consequently, MNHN 792 is a junior objective synonym of *Vespertilio molossus* Pallas, 1766. Although the type locality was not specified precisely by Geoffroy, his statement (1805b:154) that this bat and the three preceding it on page 155 came from "l'Amerique du nord, de Surinam, et principalement de Caienne," leads me to restrict the type locality for *M. longicaudatus* to Cayenne, French Guiana. *M. molossus* from the Lesser Antilles average larger in size than specimens from the northern coast of South America, based on MANOVA results obtained in this study. With a forearm of only 36.9 millimeters for MNHN 792 (see Carter and Dolan, 1978), it seems unlikely to me that this bat came from any place other than the mainland of South America and most probably somewhere along the northern coast. As outlined above, there is little doubt that *M. longicaudatus,* a skin and skull, is bat number DCDVII of Daubenton. As such, Daubenton's description of this specimen as being in fluid together with DCDVI and not a skin may be considered a *lapsus.* In my opinion, Daubenton merely confused the mode of preparation of DCDVII with that of DCDVIII when he wrote these two rather terse accounts. The former he described as in spirits, the latter as "dried out" (*dessechee*).

The matter of applying species names to the remaining two specimens at the Paris Museum is more complicated and may never be resolved to everyone's total satisfaction. Daubenton referred to three bats, DCDVI, DCDVII, and DCDVIII as probably being all of the same species. DCDVII has been identified above as the holotype of *M. longicaudatus.* Of the remaining two bats, DCDVI is the one shown in figure 1, plate XIX (Buffon and Daubenton, 1763), and thus is the holotype for *Vespertilio molossus major* Kerr 1792 and the lectotype of *Vespertilio molossus* Pallas, 1766 (see Husson, 1962, for this designation). The actual specimen on which this figure was based had been thought lost (Carter and Dolan, 1978). Forearm measurements given by Daubenton in the text for CDCVI (39.8 mm. = 1 pouce, 5 lignes; see Hershkovitz, 1975, for conversion equivalents) closely match those reflected in the figure (39 mm.) and are almost exactly those recorded by Carter and Dolan (1978) for specimen MNHN A.419/225 (39.9 mm.), a male syntype of *M. obscurus* Geoffroy. Geoffroy apparently had two specimens in hand when he named *M. obscurus,* MNHN A.419/225 and A.419/225*a,* but gave measurements for only one. These two bats differ in size with 225*a* being slightly larger than 225 (forearm of the former is 41.0 mm., that of the latter is 39.9). If Daubenton's measurements can be considered accurate, then I propose that his bat number DCDVI shown in figure 1 of plate XIX is the same specimen as A.419/225, which Geoffroy named *obscurus.* Although it cannot be proven, it appears as though the measurements given by Geoffroy on page 155 for *M. obscurus* were those of the larger specimen (head and body 60 mm. = 2 pouce, 2 lignes) if one assumes the head and body length of 55 mm. (2 pouce) reported by Daubenton (page 85) is as accurate as his measurement of the forearm. It is extremely curious that Geoffroy should recognize his specimen of *M. longicaudatus* as being one of Daubenton's yet fail to make the same connection with the two specimens of *obscurus.* Some important documentation defining this association might have been lost when the Royal Cabinet, of which Daubenton's specimens were a part, was transported from Versailles to Paris as a consequence of the French Revolution. Whatever the reason, Geoffroy apparently did not think he had either of the larger *molossus* described by Daubenton and as such gave the two fluid-preserved specimens in his charge a new name.

The name *Molossus fusciventer* Geoffroy has been considered by all taxonomists to date as a junior objective synonym of *Vespertilio molossus major* Kerr, that is, that it too refers to plate XIX (fig. 1) of Buffon and Daubenton (1763). Geoffroy made it perfectly clear (p. 155) that he did not actually have a specimen when he described *M. fusciventer* and that the name was indeed being applied to one of Daubenton's specimens. However, rather than bat number DCDVI, I believe Geoffroy actually assigned the name *fusciventer* to DCDVIII. My reasons for thinking so are as follows. Daubenton specifically gave the name "mulot volant" to DCDVI and to this bat only but did so in a footnote that appears at the end of a three-and-a-half page description of that specimen, which places the notation directly beneath the account for DCDVII. Geoffroy appears to have mistaken the reference as applying to DCDVII. This would explain why in the account for *Molossus longicaudatus* Geoffroy claims this bat was "described by M. Daubenton under the name *mulot volant.* . . ." It also explains the rather curious introduction to Geoffroy's account for *M. fusciventer* wherein he referred to that species as "le second mulot volant que M. Daubenton a decrit dans l'Histoire generale et particuliere." Knowing that Daubenton had three bats numbered DCDVI, DCDVII, and DCDVIII, all of which he thought of as conspecific and which thus could all be considered specimens of the bat "mulot volant," the "second" would necessarily be DCDVII, but Geoffroy had already recognized that specimen under the name *longicaudatus*. If Geoffroy thought the vernacular name "mulot volant" was applied to DCDVII, the "second" one would be DCDVIII. I find it odd that the measurements Geoffroy quotes for *fusciventer* are precisely those given by Daubenton for DCDVI and suspect Geoffroy lifted these numbers from Daubenton's text and used them for *fusciventer*. After all, Daubenton (p. 88) thought DCDVIII and DCDVI were of the same species and even remarked that they were more like one another than either resembled DCDVII so that body measurements, in Geoffroy's mind, might apply equally to such similar specimens. The specific epithet *fusciventer* thus would not be a junior objective synonym of *Molossus obscurus* Geoffroy and specifically apply to specimen MNHN A.419/225.

To recapitulate, three specimens of small *Molossus* are extant in the Paris Museum. MNHN 792 is specimen DCDVII of Daubenton and the holotype for *M. longicaudatus* Geoffroy and *Vespertilio molossus minor* Kerr, and the paralectotype for *Vespertilio molossus* Pallas. MNHN A.419/225 is bat number DCDVI of Daubenton and the holotype for *Vespertilio molossus major* Kerr and the lectotype for *Vespertilio molossus* Pallas; it is also a syntype of *Molossus obscurus* Geoffroy. MNHN A.419/225*a* is the same as number DCDVIII of Daubenton and the holotype for *Molossus fusciventer* Geoffroy as well as syntype of *Molossus obscurus* Geoffroy.

Overall, four major clusters are identifiable in the *M. molossus* complex, arranged here in order of increasing size (refer to Figs. 2 and 3): specimens from Panamá and the northern coast of South America, including portions of Ecuador; Lesser Antilles; Central America; South American mainland and

Greater Antilles. Representatives from the Greater Antilles average larger than both Middle American and Lesser Antillean mastiff bats and are on almost the same order of size as *molossus* from Ecuador and Perú. Electrophoretic data attest to the genetic similarity of the first three groups (see Fig. 5 and Table 7), and for this reason all have been assigned herein to the same species. Should additional collecting confirm a persistent morphological difference worthy of subspecific recognition between Lesser Antillean mastiff bats and those on the northern coast of the South American mainland, the former should be referred to *M. m. molossus;* the latter would be *M. m. minor* Kerr, 1792, with *M. longicaudatus* Geoffroy, 1805, *M. pygmaeus* Miller, 1900, and *M. daulensis* J. A. Allen, 1916, as junior synonyms. No subspecific epithet currently is available for Central American populations. Without genic information on the larger bats from Ecuador, Perú, and Argentina it is impossible to comment on their precise taxonomic position. Because of the strong physical resemblance they bear to other *M. molossus*-like bats, I place them in that species but recognize the subspecies *crassicaudatus* in deference to their larger size. *M. obscurus currentium* Thomas, 1901, would be a junior synonym. A few bats from Surinam have been examined but it is difficult to assign them to a particular subspecies inasmuch as they are intermediate in size between specimens from the Lesser Antilles and those from the eastern Andes.

Inadequate comparative material from the Greater Antilles also has hampered unravelling the association between these and mainland populations. Morphologically, they clearly fall within the *M. molossus* group, hence their inclusion in the foregoing synonomy, but on the average are larger than their geographically closest allies, Central American and Lesser Antillean populations (see Figs. 2 and 3). However, the occurrence in Jamaican populations of an otherwise *sinaloae* species-specific EST-2 108 allele points to the need for further genetic sampling in the region before a clear understanding of the specific status and phylogenetic origins of these bats can be achieved. If the *Molossus* from the Greater Antilles are found to be specifically distinct or at least deserving of subspecific recognition, the oldest available name would be *tropidorhynchus* Gray, 1839. *M. fuliginosus* Gray, 1838, although antedating *tropidorhynchus,* is unavailable due to its preoccupation by *M. fuliginosus* Cooper, 1837, and was assigned the replacement name *milleri* by Johnson (1952)

A southern invasion route for *Molossus* into the Lesser Antilles has been postulated by Koopman (1968). Besides a reaffirmation here of the strong morphological similarity noted by Koopman between the island bats and representatives from the north coast of South America and Panamá (compare populations 59–63 to 51, 52, 46, 79 in Figs. 2 and 3) that makes this assumption plausible, the near fixation of EST-2 116 in Dominican and Venezuelan populations further strengthens that assertion by providing a genetic link as well. Although Koopman (1975) also regarded the Lesser Antilles as the source of colonizers for the Greater Antilles, Baker and Genoways (1978) later considered the issue still unresolved because of the equal likelihood of dispersal to Cuba or Jamaica from the Yucatán Peninsula or adjacent parts of Middle

America. Koopman (1975) discerned a difference between the small *molossus* on Puerto Rico and the Virgin Islands and those inhabiting the northern Lesser Antilles, referring the former and larger to *M. m. fortis* and the latter to *M. m. debilis* (the holotype of which was taken on St. Kitts). The line of demarcation between the two subspecies was fixed along the Anegada Passage. In light of my own findings of a similar morphological hiatus and the presence of a seemingly anomalous *sinaloae* EST-2 108 allele, at least in Jamaican mastiffs, I am inclined to propose a separate invasion of the Greater Antilles from Middle America rather than accept the Greater Antillean fauna as an attenuation of a Lesser Antillean distribution of small *molossus*.

The only record known to me of this taxon on the Caribbean versant in Middle America south of México is that of Murie (1935), who reported obtaining the remains of a "Molossus aztecus" from a falcon in Guatemala. I have assigned this specimen to *M. molossus* rather than *M. aztecus* based on the relatively low elevation (approximately 1500 feet) and dry, mountain pine ridge habitat in which the falcon was shot, and have included it here because of its potential import in clarifying the distribution of this species. However, the otherwise apparent absence of *M. molossus* along the Caribbean versant (see Figs. 12 and 15), at least in Recent times, raises curious questions regarding the ancestral origins of the small *M. molossus* inhabiting the Greater Antilles. A remote link between those populations and *M. sinaloae*, which is widespread on the Caribbean side, should not be overlooked by future researchers.

Specimens examined (263).—Populations 33, 34, 36–38, 40 (=44), 41, 42, 46, 51–54, 57, 59–66, 73–75, 78, 79, 81, 85 (see Table 1). EL SALVADOR. Cuscatlán: Suchitoto, 3 (TCWC). San Salvador: 1 mi. NW San Salvador, 2 (KU). MÉXICO. Chiapas: Huehuetán, 5 (USNM); 1 mi. SE Puerto Madero, 1 (KU). NICARAGUA. Boaco: 17 km. N, 15 km. E Boaco, 7 (KU). Chinandega: Potosí, 1 (KU); El Realejo, Hacienda San Isidro, 1 (USNM). Managua: 3 mi. SW Managua, 8 (KU). Rivas: 12.5 mi. S, 3 mi. E Rivas, 2 (TCWC). SURINAM: Paramaribo, 4 (TCWC).

Additional records.—HONDURAS. La Paz: El Monteado; Los Encuentros (Goodwin, 1942)—although described as *bondae*, LaVal (1977) concluded after examining these specimens that they were not that species; forearm measurements indicate they represent *molossus*. MÉXICO. Coahuila: Morelos (Ramírez-Pulido and López-Forment, 1979). Colima: Pueblo Juárez. Tamaulipas: Rancho "La Isla," 3 km. N El Limón (Villa-R., 1967).

Acknowledgments

To Dilford C. Carter, committee chairman, I extend my warmest thanks for his guidance and support during my academic tenure. Robert J. Baker graciously provided laboratory space, support equipment, and perhaps most importantly, a stimulating environment in which to work. Other members of my committee for whose counsel over the years I am most grateful are J. Knox Jones, Jr., Raymond C. Jackson, Ronald K. Chesser, and the late Robert L. Packard. T. E. Brady and J. S. Sevall lent invaluable assistance in the isoelectric analysis. Curators who generously loaned specimens and made critical comparisons to type specimens in their care deserve thanks and recognition: H. H. Genoways (Carnegie Museum of Natural History); K. F. Koopman (American Museum of Natural History); D. R. Patten (Los Angeles County Natural History Museum); D. J. Schmidly (Texas Cooperative Wildlife Collection, Texas A&M University); D. E. Wilson (U.S. National Museum). Several individuals supplied specimens from regions in which I was unable to collect and which were important in clarifying systematic relationships. Among them are: J. Bowles and J. Groen, Yucatán; P. V. and L. August and R. J. Baker, Venezuela; H. H. Genoways, R. J. Baker, and J. W. Bickham, Jamaica (M. Graham Netting Research Fund through a grant from the S. May Charitable Trust); J. K. Jones, Jr., and R. J. Baker, Dominica. R. K. Chesser made computer programs available for portions of the genetic analysis.

Many fellow graduate students over the years contributed much to my understanding of biology and I owe them a special thanks for their criticism, suggestions, and friendship: M. Arnold, R. Barnett, B. Bass, P. O'Connell, I. Greenbaum, H. Haiduk, R. Honeycutt, A. Johnson, B. Koop, Lynn and Laurie Robbins, T. Yates.

This study was supported in part by grants from The Society of the Sigma Xi.

LITERATURE CITED

AELLEN, V. 1970. Catalogue raisonné des chiroptères de la Colombie. Rev. Suisse Zool., 77: 1–37.

ALLEN, J. A. 1906. Mammals from the states of Sinaloa and Jalisco, Mexico, collected by J. H. Batty during 1904 and 1905. Bull. Amer. Mus. Nat. Hist., 22:191–262.

———. 1916. List of mammals collected in Colombia by the American Museum of Natural History expeditions, 1910–1915. Bull. Amer. Mus. Nat. Hist., 35:191–238.

ALVAREZ, T., AND J. RAMÍREZ-PULIDO. 1972. Notas acerca de murciélagos mexicanos. An. Esc. Nac. Cienc. Biol., México, 19:167–178.

BAKER, R. J., AND R. A. BASS. 1979. Evolutionary relationship of the Brachyphyllinae to the Glossophagine genera *Glossophaga* and *Monophyllus*. J. Mamm., 60:364–372.

BAKER, R. J., AND J. W. BICKHAM. 1980. Karyotypic evolution in bats: Evidence of extensive and conservative chromosomal evolution in closely related taxa. Syst. Zool., 29: 239–253.

BAKER, R. J., AND H. H. GENOWAYS. 1978. Zoogeography of Antillean bats. Pp. 53–97, *in* Zoogeography in the Caribbean (F. B. Gill, ed.), Spec. Publ. Acad. Nat. Sci. Philadelphia, 13:iii+1–128.

BARR, A. J., J. H. GOODNIGHT, J. P. SALL, AND J. T. HELWIG. 1976. A user's guide to SAS76. SAS Institute, Raleigh, North Carolina, 329 pp.

BENGTSSON, B. O. 1978. Avoid inbreeding: at what cost? J. Theor. Biol., 73:439–444.

BICKHAM, J. W., AND R. J. BAKER. 1979. Canalization model of chromosomal evolution. Pp. 70–84, *in* Models and methodologies in evolutionary theory (J. H. Schwartz and H. B. Rollins, eds.), Bull. Carnegie Mus., 13:1–105.

BUFFON, [G. L. LE CLERC COMTE] DE, AND [L. J. M.] DAUBENTON. 1763. Histoire naturelle, générale et particuliére, avec la description du Cabinet du Roi. Paris, 10:xii+368 pp.

BUSH, G. L., S. M. CASE, A. C. WILSON, AND J. L. PATTON. 1977. Rapid speciation and chromosomal evolution in mammals. Proc. Nat. Acad. Sci. Philadelphia, 74:3942–3946.

CARTER, D. C. 1962. The systematic status of the bat *Tadarida brasiliensis* (I. Geoffroy) and its related mainland forms. Unpublished Ph.D. dissertation, Texas A&M Univ., College Station, 80 pp.

———. 1970. Chiropteran reproduction. Pp. 233–246, *in* About bats (B. H. Slaughter and D. W. Walton, eds.), Southern Methodist Univ. Press, Dallas, vii+1–339 pp.

CARTER, D. C., AND P. G. DOLAN. 1978. Catalogue of type specimens of Neotropical bats in selected European museums. Spec. Publ. Mus., Texas Tech Univ., 15:1–136.

CHESSER, R. K. 1983. Genetic variability within and among populations of the black-tailed prairie dog. Evolution, 37:320–331.

DALQUEST, W. W. 1954. Netting bats in tropical Mexico. Trans. Kansas Acad. Sci., 57:1–10.

DAVIS, W. B. 1951. Bat, *Molossus nigricans*, eaten by the rat snake, *Elaphe laeta*. J. Mamm., 32:219.

DAVIS, W. B., AND R. J. RUSSELL. 1954. Mammals of the Mexican state of Morelos. J. Mamm., 35:63–80.

DOLAN, P. G., AND D. C. CARTER. 1979. Distributional notes and records for Middle American Chiroptera. J. Mamm., 60:644–649.

DUELLMAN, W. E. 1960. A distributional study of the amphibians of the Isthmus of Tehuantepec, México. Univ. Kansas Publ., Mus. Nat. Hist., 13:19–72.

———. 1966. The Central American herpetofauna: an ecological perspective. Copeia, 4: 700–719.

DULIĆ, B., AND M. MRAKOVČIĆ. 1980. Chromosomes of European free-tailed bat, Tadarida teniotis teniotis (Rafinesque, 1814, Mammalia, Chiroptera, Molossidae). Biosistematika, 6:109–112.

DUNN, E. R. 1940. Some aspects of herpetology in lower Central America. Trans. New York Acad. Sci., 2:156–158.

ECHELLE, A. A., A. F. ECHELLE, AND B. A. TABER. 1976. Biochemical evidence for congeneric

67

competition as a factor restricting gene flow between populations of a darter (Percidae: Etheostoma). Syst. Zool., 25:228–235.

EGER, J. L. 1977. Systematics of the genus *Eumops* (Chiroptera: Molossidae). Life Sci. Contrib., Royal Ontario Mus., 110:1–69+iv.

FLEMING, T. H. 1971. Population ecology of three species of Neotropical rodents. Misc. Publ. Mus. Zool., Univ. Michigan, 143:1–77.

FREEMAN, P. W. 1981. A multivariate study of the family Molossidae (Mammalia, Chiroptera): morphology, ecology, evolution. Fieldiana Zool., n.s., 7:vii+1–173.

GARDNER, A. L. 1966. A new subspecies of the Aztec mastiff bat, *Molossus aztecus* Saussure, from southern Mexico. Contrib. Sci., Los Angeles Co. Mus. Nat. Hist., 111:1–5.

GARDNER, A. L., R. K. LAVAL, AND D. E. WILSON. 1970. The distributional status of some Costa Rican bats. J. Mamm., 51:712–729.

GENOWAYS, H. H., R. C. DOWLER, AND C. H. CARTER. 1981. Intraisland and interisland variation in Antillean populations of *Molossus molossus* (Mammalia: Molossidae). Ann. Carnegie Mus., 50:475–492.

GEOFFROY ST.-HILAIRE, É. 1805a. Note sur une petite famille de chauve-souris d'Amérique, désignée sous le nom générique de Molossus. Bull. Sci. Soc. Philom., Paris, 3(96):278–279 (error pro 378–379).

———. 1805b. Mémoire sur quelques chauve-souris d'Amérique formant une petite famille sous le nom molossus. Ann. Mus. Nat. Hist., Paris, 6:150–156.

GOODWIN, G. G. 1942. Mammals of Honduras. Bull. Amer. Mus. Nat. Hist., 79:107–195.

———. 1946. Mammals of Costa Rica. Bull. Amer. Mus. Nat. Hist., 87:271–473.

———. 1956. A preliminary report on the mammals collected by Thomas MacDougall in southeastern Oaxaca, Mexico. Amer. Mus. Novit., 1757:1–15.

———. 1959. Descriptions of some new mammals. Amer. Mus. Novit., 1967:1–8.

———. 1960. The status of *Vespertilio auripendulus* Shaw, 1800, and *Molossus ater* Geoffroy, 1805. Amer. Mus. Novit., 1944:1–6.

GOODWIN, G. G., AND A. M. GREENHALL. 1961. A review of the bats of Trinidad and Tobago. Bull. Amer. Mus. Nat. Hist., 122:187–301 + 7–46 pls.

GREENBAUM, I. F. 1978. Evolutionary genetics and speciation of the tent-making bat *Uroderma* (Chiroptera: Phyllostomatidae). Unpublished Ph.D. dissertation, Texas Tech Univ., Lubbock, 96 pp.

GREENBAUM, I. F., R. J. BAKER, AND J. H. BOWERS. 1978. Chromosomal homology and divergence between sibling species of deer mice: *Peromyscus maniculatus* and *P. melanotis*. Evolution, 32:334–341.

HAFFER, J. 1967a. Speciation in Colombian forest birds west of the Andes. Amer. Mus. Novit., 2294:1–57.

———. 1967b. Zoogeographical notes on the "nonforest" lowland birds of northwestern South America. El Hornero (Buenos Aires), 10:315–333.

HANDLEY, C. O., JR. 1966. Checklist of the mammals of Panama. Pp. 753–795, *in* Ectoparasites of Panama (R. L. Wenzel and V. J. Tipton, eds.), Field Mus. Nat. Hist., Chicago, xii+861 pp.

———. 1976. Mammals of the Smithsonian Venezuelan project. Brigham Young Univ. Sci. Bull., 20:1–89 + map and gazeteer.

HEDGECOCK, D. 1978. Population subdivision and genetic divergence in the red-bellied newt, *Taricha rivularis*. Evolution, 32:271–286.

HONEYCUTT, R. L., AND D. J. SCHMIDLY. 1979. Chromosomal and morphological variation in the Plains pocket gopher, Geomys bursarius, in Texas and adjacent states. Occas. Papers Mus., Texas Tech Univ., 58:1–54.

HUSSON, A. M. 1962. The bats of Suriname. Zool. Verhand., Leiden, 58:1–282 + 30 pls.

JOHNSON, D. H. 1952. A new name for the Jamaican bat *Molossus fuliginosus* Gray. Proc. Biol. Soc. Washington, 65:197–198.

JONES, J. K., JR. 1966. Bats from Guatemala. Univ. Kansas Publ., Mus. Nat. Hist., 16:439–472.

JONES, J. K., JR., J. D. SMITH, AND R. W. TURNER. 1971. Noteworthy records of bats from Nicaragua, with a check list of the chiropteran fauna of the country. Occas. Papers Mus. Nat. Hist., Univ. Kansas. 2:1–35.

KOOPMAN, K. F. 1968. Taxonomic and distributional notes on Lesser Antillean bats. Amer. Mus. Novit., 2333:1–13.

———. 1970. Zoogeography of bats. Pp. 29–50, *in* About bats (B. H. Slaughter and D. W. Walton, eds.), Southern Methodist Univ. Press, Dallas, vii+1–339 pp.

———. 1975. Bats of the Virgin Islands in relation to those of the Greater and Lesser Antilles. Amer. Mus. Novit., 2581:1–7.

———. 1978. Zoogeography of Peruvian bats with special emphasis on the role of the Andes. Amer. Mus. Novit., 2651:1–33.

LANDE, R. 1979. Effective deme sizes during long-term evolution estimated from rates of chromosomal rearrangement. Evolution, 33:234–251.

LARSON, A., AND R. HIGHTON. 1979. Geographic protein variation and divergence in the salamanders of the *Plethodon welleri* group (Amphibia, Plethodontidae). Syst. Zool., 27:431–448.

LAVAL, R. K. 1977. Notes on some Costa Rican bats. Brenesia (Museo Nacional de Costa Rica), 10/11:77–83.

LAVAL, R. K., AND H. S. FITCH. 1977. Structure, movements and reproduction in three Costa Rican bat communities. Occas. Papers Mus. Nat. Hist., Univ. Kansas, 69:1–28.

LEOPOLD, A. S. 1952. Zonas de vegetación en México. Bol. Soc. Mexicana de Geog. y Estad., 73:47–93 + 1 map.

LONG, C. A. 1968. An analysis of patterns of variation in some representative Mammalia. Part I. A review of estimates of variability in selected measurements. Trans. Kansas Acad. Sci., 71:201–227.

———. 1969. An analysis of patterns of variation in some representative Mammalia. Part II. Studies on the nature and correlation of measures of variation. Misc. Publ. Mus. Nat. Hist., Univ. Kansas, 51:289–302.

LONG, C. A., AND C. J. JONES. 1966. Variation and frequency of occurrence of the baculum in a population of Mexican free-tailed bats. Southwestern Nat., 11:290–295.

LOUDENSLAGER, E. J., AND G. A. E. GALL. 1980. Geographic patterns of protein variation and subspeciation in cutthroat trout, *Salmo clarki*. Syst. Zool., 29:27–42.

MARINKELLE, C. J., AND A. CADENA. 1972. Notes on bats new to the fauna of Colombia. Mammalia, 36:50–58.

MCCRACKEN, G. F., AND J. W. BRADBURY. 1977. Paternity and genetic heterogeneity in the polygynous bat *Phyllostomus hastatus*. Science, 198:303–306.

MILLER, G. S. 1902. Twenty new American bats. Proc. Acad. Nat. Sci. Philadelphia, 54:389–412.

———. 1907. The families and genera of bats. Bull. U.S. Nat. Mus., 57:xvii+1–1282.

———. 1913. Notes on the bats of the genus Molossus. Proc. U.S. Nat. Mus., 46:85–92.

MURIE, A. 1935. Mammals from Guatemala and British Honduras. Misc. Publ. Mus. Zool., Univ. Michigan, 26:1–30.

NEI, M. 1972. Genetic distance between populations. Amer. Nat., 106:283–292.

———. 1973. Analysis of gene diversity in subdivided populations. Proc. Nat. Acad. Sci. Philadelphia, 70:3321–3323.

———. 1977. *F*-statistics and analysis of gene diversity in subdivided populations. Ann. Hum. Genet., London, 41:225–233.

OJASTI, J., AND O. J. LINARES. 1971. Adiciones a la fauna de murciélagos de Venezuela con notas sobre las especies del genero *Diclidurus* (Chiroptera). Acta Biol. Venezuela, 7:421–441.

PALLAS, P. S. 1766. Miscellanea zoologica quibus novae imprimis atque obscurae animalium species describunter et observationibus inconibusque illustratum. Hague Comitum, xii+224 pp. + xiv pls.

PATTON, J. L. 1967. Chromosomal studies of certain pocket mice, genus *Perognathus* (Rodentia: Heteromyidae). J. Mamm., 48:27–37.

PATTON, J. L., AND S. Y. YANG. 1977. Genetic variation in *Thomomys bottae* pocket gophers: macrogeographic patterns. Evolution, 31:697–720.

PETERSON, R. L. 1971. The systematic status of the African molossid bats *Tadarida bemmeleni* and *Tadarida cistura*. Canadian J. Zool., 49:1347–1354.

————. 1974. Variation in the African bat, *Tadarida lobata*, with notes on habitat and habits (Chiroptera: Molossidae). Life Sci. Occas. Papers, Royal Ontario Mus., 24:1–8.

PETERSON, R. L., AND D. W. NAGORSEN. 1975. Chromosomes of fifteen species of bats (Chiroptera) from Kenya and Rhodesia. Life Sci. Occas. Papers, Royal Ontario Mus., 27:1–14.

PREGILL, G. K., AND S. L. OLSON. 1981. Zoogeography of West Indian vertebrates in relation to Pleistocene climatic cycles. Ann. Rev. Ecol. Syst., 12:75–98.

RAMÍREZ-PULIDO, J., AND W. LÓPEZ-FORMENT. 1979. Additional records of some Mexican bats. Southwestern Nat., 24:541–543.

ROGERS, J. S. 1972. Biochemical polymorphism and systematics in the genus *Peromyscus*. IV. Measures of genetic similarity and genetic distance. Stud. Genetics VII, Univ. Texas Publ., 7213:145–153.

ROHLF, F. J., AND J. KISHPAUGH. 1972. Numerical taxonomy system of multivariate statistical programs. The State Univ. of New York at Stoney Brook, 87 pp.

RYMAN, N., C. REUTERWALL, K. NYGREN, AND T. NYGREN. 1980. Genetic variation and differentiation in Scandanavian moose (*Alces alces*): are large mammals monomorphic? Evolution, 34:1037–1049.

SCHOENER, T. W. 1965. The evolution of bill size differences among sympatric congeneric species of birds. Evolution, 19:189–213.

————. 1968. The *Anolis* lizards of Bimini: resource partitioning in a complex fauna. Ecology, 49:704–726.

SAUSSURE, H. DE. 1860. Note sur quelques Mammifères du Mexique. Rev. Mag. Zool., Paris, ser. 2, 12:281–293.

SCHWARTZ, O. A., AND K. B. ARMITAGE. 1980. Genetic variation in social mammals: the marmot model. Science, 207:665–667.

SELANDER, R. K. 1970. Behavior and genetic variation in natural populations. Amer. Zool., 10:53–66.

SELANDER, R. K., AND D. W. KAUFMAN. 1975. Genetic structure of populations of the brown snail (*Helix aspersa*). I. Microgeographic variation. Evolution, 29:385–401.

SELANDER, R. K., M. H. SMITH, S. Y. YANG, W. E. JOHNSON, AND J. B. GENTRY. 1971. Biochemical polymorphism and systematics in the genus *Peromyscus*. I. Variation in the old-field mouse (*Peromyscus polionotus*). Stud. Genetics VI, Univ. Texas Publ., 7103:49–90.

SLATKIN, M. 1976. The rate of spread of an advantageous allele in a subdivided population. Pp. 767–780, *in* Population genetics and ecology (S. Karlin and E. Nevo eds.), Academic Press, New York.

SMITH, J. D., AND A. STARRETT. 1979. Morphometric analysis of chiropteran wings. Pp. 229–316, *in* Biology of bats of the New World family Phyllostomatidae. Part III (R. J. Baker, J. Knox Jones, Jr., and D. C. Carter, eds.), Spec. Publ. Mus., Texas Tech Univ., 16:1–441.

TSCHUDI, J. J. VON. 1844. Therologie. *In* Untersuchungen über die Fauna Peruana. St. Gallen, pt. 1, pp. 1–262.

TUTTLE, M. D. 1970. Distribution and zoogeography of Peruvian bats, with comments on natural history. Univ. Kansas Sci. Bull., 49:45–86.

ULARGUI, C. J. I. 1979. Biología y ecología de los murciélagos del hato "El Frio," Apura, Venezuela. Unpublished Ph.D. dissertation, Universidad Politecnica de Madrid, ix+401 pp.

VALDIVIESO, D. 1964. La fauna quiróptera del Departamento de Cundinamarca, Colombia. Rev. Biol. Trop., 12:19–45.

VAUGHAN, T. A. 1978. Mammalogy. W. B. Saunders, Philadelphia, x+522 pp.

VILLA-R., B. 1967. Los murciélagos de México. Inst. Biol., Univ. Nac. Autónoma México, xvi+491 pp. [Published 6 February 1967.]

WARNER, J. W., J. L. PATTON, A. L. GARDNER, AND R. J. BAKER. 1974. Karyotypic analyses of twenty-one species of molossid bats (Molossidae: Chiroptera). Canadian J. Genet. Cytol., 16:165–176.

WATKINS, L. C., J. K. JONES, JR., AND H. H. GENOWAYS. 1972. Bats of Jalisco, México. Spec. Publ. Mus., Texas Tech Univ., 1:1–44.

WILSON, A. C., G. L. BUSH, S. M. CASE, AND C. M. KING. 1975. Social structuring of mammalian populations and rate of chromosomal evolution. Proc. Nat. Acad. Sci. Philadelphia, 72:5061–5035.

WRIGHT, S. 1965. The interpretation of population structure by *F*-statistics with special regard to systems of mating. Evolution, 9:395–420.

———. 1978. Evolution and the genetics of populations. Vol. 4. Variability within and among natural populations. Univ. Chicago Press, 580 pp.

———. 1980. Genic and organismic selection. Evolution, 34:825–843.

YATES, T. L. 1978. The systematics and evolution of North American moles (Insectivora: Talpidae). Unpublished Ph.D. dissertation, Texas Tech Univ., Lubbock, xi+304 pp.

YATES, T. L., AND D. J. SCHMIDLY. 1977. Systematics of Scalopus aquaticus (Linnaeus) in Texas and adjacent states. Occas. Papers Mus., Texas Tech Univ., 45:1–36.

Address of author: *c/o The Museum, Texas Tech University, Lubbock, Texas 79409. Received 24 March, accepted 4 April 1988.*